"十二五"普通高等教育规划教材

SHIPIN ZHUANYE YINGYU
食品专业英语

李向阳　张建友　主编

中国质检出版社
中国标准出版社
北京

图书在版编目(CIP)数据

食品专业英语/李向阳,张建友主编. —北京:中国质检出版社,2013(2020.1 重印)
"十二五"普通高等教育规划教材
ISBN 978-7-5026-3737-8

Ⅰ.①食… Ⅱ.①李… ②张… Ⅲ.①食品专业英语-高等教育-教材
Ⅳ.①TS201.2

中国版本图书馆 CIP 数据核字(2012)第 290968 号

内 容 提 要

本书依据食品专业学生的实际学习需要,系统地选编了食品安全质量管理、食品科学基础、食品工艺学等方面的内容。本书突出了食品安全和质量管理的重要性,但食品安全与质量管理是建立在食品化学和食品工艺学基础之上的,因此,在第二章介绍了食品化学与营养的相关知识,第三章介绍了食品工艺学的部分知识。

本书可作为高等学校食品及相关专业的英语教材,亦可供食品管理、食品生产企业相关人员学习和参考。

中国质检出版社
中国标准出版社 出版发行
北京市朝阳区和平里西街甲 2 号(100029)
北京市西城区三里河北街 16 号(100045)
网址 www.spc.net.cn
总编室:(010)68533533 发行中心:(010)51780238
读者服务部:(010)68523946
中国标准出版社秦皇岛印刷厂印刷
各地新华书店经销
*
开本 787×1092 1/16 印张 8.25 字数 155 千字
2013 年 1 月第一版 2020 年 1 月第六次印刷
*
定价:19.00 元

如有印装差错 由本社发行中心调换
版权专有 侵权必究
举报电话:(010)68510107

审定委员会

陈宗道（西南大学）

谢明勇（南昌大学）

殷涌光（吉林大学）

李云飞（上海交通大学）

何国庆（浙江大学）

王锡昌（上海海洋大学）

林　洪（中国海洋大学）

徐幸莲（南京农业大学）

侯正明（哈尔滨工业大学）

吉鹤立（上海市食品添加剂行业协会）

巢强国（上海市质量技术监督局）

本书编委会

主　编　李向阳（山东农业大学）

　　　　张建友（浙江工业大学）

副主编　包　斌（上海海洋大学）

　　　　刁恩杰（山东农业大学）

　　　　李慧静（河北农业大学）

编　委　李德海（东北林业大学）

　　　　张英春（哈尔滨工业大学）

序　言

近年来，人们对食品安全的关注度日益增强，食品行业已成为支撑国民经济的重要产业和社会的敏感领域。随着食品产业的进一步发展，食品安全问题层出不穷，对整个社会的发展造成了一定的不利影响。为了保障食品安全，规制食品产业的有序发展，近期国家对食品安全的监管和整治力度不断加强。经过各相关主管部门的不懈努力，我国已基本形成并明确了卫生与农业部门实施食品原材料监管、质监部门承担食品生产环节监管、工商部门从事食品流通环节监管的制度完善的食品安全监管体系。

在整个食品行业快速发展的同时，行业自身的结构性调整也在不断深化，这种调整使其对本行业的技术水平、知识结构和人才特点提出了更高的要求，而与此相关的高等教育正是在食品科学与工程各项理论的实际应用层面培养专业人才的重要渠道。因此，近年来教育部对食品类各专业的高等教育发展日益重视，并连年加大投入以提高教育质量，以期向社会提供更加适应经济发展的应用型技术人才。为此，教育部对高等院校食品类各专业的具体设置和教材目录也多次进行了相应的调整，使高等教育逐步从偏重基础理论的教育模式中脱离出来，使其真正成为为国家培养应用型的高级技术人才的专业教育，"十二五"期间，这种转化将加速推进并最终得以完善。为适应这一特点，编写高等院校食品类各专业所需的教材势在必行。

针对以上变化与调整，由中国质检出版社牵头组织了"十二五"普通高等教育规划教材（食品类）的编写与出版工作，该套教材主要适用于高等院校的食品类各相关专业。由于该领域各专业的技术应用性强、知识结构更新快，因此，我们有针对性地组织了西南大学、南昌大学、上海交通大学、浙江大学、上海海洋大学、中国海洋大学、南京农业大学、华中农业大学、浙江工业大学以及河北农业大学等40多所相关高校、科研院所以及行业协会中兼具丰富工程实践和教学经验的专家学者担当各教材的主编与主审，从而为我们成功推出这套框架好、内容新、适应面广的好教材提供了必要的保障，以此来满足食品类各专业普通高等教育的不断发展和当前全社会范围内对建立食品安全体系的迫切需要；这也对培养素质全面、适应性强、有创新能力的应用型技术人才，进一步提高食品类各专业高等教育教材的编写水平起到了积极的推动作用。

针对应用型人才培养院校食品类各专业的实际教学需要，本系列教材的编写尤其注重了理论与实践的深度融合，不仅将食品科学与工程领域科技发展的新理论合理融入教材中，使读者通过对教材的学习，可以深入把握食品行业发展的全貌，而且也将食品行业的新知识、新技术、新工艺、新材料编入教材中，使读者掌握最先进的知识和技能，这对我国新世纪应用型人才的培养大有裨益。相信该套教材的成功推出，必将会推动我国食品类高等教育教材体系建设的逐步完善和不断发展，从而对国家的新世纪人才培养战略起到积极的促进作用。

教材审定委员会

2012 年 12 月

前 言
• FOREWORD •

食品工业是我国国民经济的重要支柱产业。无论是食品的国际贸易还是食品专业技术的发展，都离不开国际交流。尤其是加入WTO以来，国际间的信息交流异常活跃。因此，掌握食品专业英语及相关知识，学会与国外同行进行技术交流与沟通是食品专业学生的一项基本技能。

本书系统地选编了食品安全质量管理、食品科学基础、食品工艺学等方面的内容。加强食品安全与质量管理，提高食品的安全和质量是目前各国政府、企业和个人特别关注的事情。因此，本书突出了食品安全和质量管理的重要性，但食品安全与质量管理是建立在食品化学和食品工艺学基础之上的，了解其基础知识也是相当重要的，因此第二章介绍了食品化学与营养的相关知识，第三章介绍了食品工艺学的部分知识。

本书内容具有较强的实用性和指导性，可为食品管理者、企业和食品专业的学生提供理论指导。通过本书的学习，可使学生们提高灵活运用与食品相关的词汇和用语以及在食品听、说、读、写等方面的能力，为以后的进一步学习或就业打下良好的英语基础。

由于资料收集和撰写水平有限，不妥之处，敬请读者指正。

编　者

2012 年 11 月

目 录
• CONTENTS •

Chapter 1　Food Safety and Management ⋯⋯⋯⋯⋯⋯⋯⋯⋯⋯⋯⋯（1）

　　Unit 1　Overview of Food Safety ⋯⋯⋯⋯⋯⋯⋯⋯⋯⋯⋯⋯⋯⋯⋯（1）

　　Unit 2　Issues in Food Safety ⋯⋯⋯⋯⋯⋯⋯⋯⋯⋯⋯⋯⋯⋯⋯⋯（8）

　　Unit 3　GMP and SSOP ⋯⋯⋯⋯⋯⋯⋯⋯⋯⋯⋯⋯⋯⋯⋯⋯⋯⋯（18）

　　Unit 4　HACCP ⋯⋯⋯⋯⋯⋯⋯⋯⋯⋯⋯⋯⋯⋯⋯⋯⋯⋯⋯⋯⋯（27）

　　Unit 5　Food Risk Analysis ⋯⋯⋯⋯⋯⋯⋯⋯⋯⋯⋯⋯⋯⋯⋯⋯（31）

　　Unit 6　Food Quality ⋯⋯⋯⋯⋯⋯⋯⋯⋯⋯⋯⋯⋯⋯⋯⋯⋯⋯⋯（38）

　　Unit 7　Food Quality Management ⋯⋯⋯⋯⋯⋯⋯⋯⋯⋯⋯⋯⋯⋯（45）

　　Unit 8　Quality Control Tools ⋯⋯⋯⋯⋯⋯⋯⋯⋯⋯⋯⋯⋯⋯⋯（50）

　　Unit 9　Crisis Management for Food Industry ⋯⋯⋯⋯⋯⋯⋯⋯⋯（59）

　　Unit 10　Foods Entry–Exit Inspection and Quarantine ⋯⋯⋯⋯⋯⋯⋯（65）

Chapter 2　Food Chemistry and Nutrition ⋯⋯⋯⋯⋯⋯⋯⋯⋯⋯⋯⋯（72）

　　Unit 1　Water ⋯⋯⋯⋯⋯⋯⋯⋯⋯⋯⋯⋯⋯⋯⋯⋯⋯⋯⋯⋯⋯（72）

　　Unit 2　Carbohydrates ⋯⋯⋯⋯⋯⋯⋯⋯⋯⋯⋯⋯⋯⋯⋯⋯⋯⋯（75）

　　Unit 3　Fats ⋯⋯⋯⋯⋯⋯⋯⋯⋯⋯⋯⋯⋯⋯⋯⋯⋯⋯⋯⋯⋯⋯（78）

　　Unit 4　Proteins ⋯⋯⋯⋯⋯⋯⋯⋯⋯⋯⋯⋯⋯⋯⋯⋯⋯⋯⋯⋯（80）

　　Unit 5　Vitamins ⋯⋯⋯⋯⋯⋯⋯⋯⋯⋯⋯⋯⋯⋯⋯⋯⋯⋯⋯⋯（83）

I

Unit 6　Minerals ··· (87)

Unit 7　Enzymes ··· (93)

Unit 8　Nutrition ··· (97)

Chapter 3　Food Technology ····································· (101)

Unit 1　Sterilized Milk and Milk Products Cultured Milk Products ·········· (101)

Unit 2　Cheese ·· (103)

Unit 3　Meat Techniques of Curing Dry Curing ····················· (107)

Unit 4　Eggs ··· (116)

References ··· (119)

Chapter 1　Food Safety and Management

Unit 1　Overview of Food Safety

1.1　Definition of Food Safety

The concept of safe and wholesome food encompasses many diverse elements. From a nutritional aspect, it is food that contains the nutrients humans need and that helps prevent long – term chronic disease, promoting health into old age. From a food safety aspect, it is food that is free not only from toxins, pesticides, and chemical and physical contaminants, but also from microbiological pathogens such as bacteria and viruses that can cause illness.

Food safety is defined as the assurance that the food will not cause harm to the consumer when it is prepared and/or eaten according to its intended use.

The term "safe food" represents different ideals to different audiences. Consumers, special interest groups, regulators, industry, and academia will have their unique descriptions based on their perspectives.

Consumers are the end users and thus are at the last link of the food supply chain from production, through processing and distribution, to retail and food service businesses. Consumers are multidimensional and multifaceted. Population differ in age, life experiences, health, knowledge, culture, sex, political views, nutritional needs, purchasing power, media inputs, family status, occupation, and education. Safe food means food that has been handled properly, including thorough washing of food that will be cooked and anything to be eaten raw. Safe food means food prepared on clean and sanitized surfaces with utensils and dishes that also are cleaned and sanitized. Other consumers want safe food that retains vitamins and minerals but does not have harmful pesticides.

Safe food is a composite of all of the views and descriptions held by consumers, special interest groups, academicians, regulatory authorities, and industry. Almost any single definition of safe food will be overly simplistic, because safe food is a complex, multifaceted concept.

1.2　Hazards Associated with Foods

A hazard is a biological, chemical or physical agent in, or condition of, food with the potential to cause an adverse health effect. All three types of hazards are associated with fresh produce comprise.

Biological hazards are composed of bacteria, parasites and viruses; **Chemical hazards** include naturally occurring hazards, added chemical hazards and contaminants; **Physical hazards** are foreign bodies like glass, wood, stones, insulation, plastic, etc.

1.2.1 Biological hazards

Biological hazards include disease-causing bacteria, viruses, and parasites. Many of microorganisms occur naturally in the environment and can be foodborne, waterborne, or transmitted from a person or an animal. Cooking kills or inactivates most pathogens, while proper cooling and storage can control them before or after cooking.

Bacteria

Bacteria are single-celled organisms so small they can only be seen with a microscope. Bacteria are everywhere and most are not pathogenic (disease-causing). The human gastrointestinal tract is home to more than 300 species of bacteria. Fortunately, only a few of these cause illness. Some bacteria are beneficial and are used in making foods such as yogurt, cheese, and beer. Others cause food to spoil, but do not cause human sickness. This difference between spoilage bacteria and pathogenic bacteria is important in the prevention of foodborne illness. Since pathogenic bacteria generally cannot be detected by looks, smell, or taste, we rely on spoilage bacteria to indicate that a food should not be eaten. Not many people will eat food that has become slimy or that smells bad. Pathogenic bacteria cause foodborne illness in three different ways:

Infection Some bacteria damage the intestines directly. This type of illness occurs from eating food contaminated with live pathogenic bacteria. Cells that are alive and reproducing are vegetative cells. Many bacteria are killed in the acidic environment of the stomach, but some survive, pass through to the small intestine, and begin to grow in number. When the bacteria have multiplied to a high enough number (this depends on the strain of bacteria, its virulence or strength, and the health and susceptibility of the individual), the person becomes ill.

Intoxication Some bacteria produce harmful toxins or other chemicals that are not present in the food. It is not the bacteria itself that causes illness, but rather the toxin the bacteria produce. This can happen even if the pathogen itself has been killed, as long as it had sufficient time to produce enough toxin before dying.

Toxico-infection Some bacteria enter the intestines live, survive the acidic environment of the stomach, and then produce a harmful toxin inside the human digestive system. Toxico-infection is a combination of the previous two examples in that live cells must be consumed, but the toxin is produced in the intestine and it is the toxin that really causes the illness.

Viruses

Several viruses also cause foodborne illness. Viruses differ from bacteria in that they are smaller, require a living animal or human host to grow and reproduce, do not multiply in foods, and are not complete cells. Ingestion of only a few viral particles is enough to produce an infection. Humans are host to a number of viruses that reproduce in the intestines and then are excreted in the feces. Thus, transmission of viruses comes from contact with sewage or water contaminated by fecal matter or direct contact with human fecal material. Human pathogenic viruses are often discharged into marine waters through treated and untreated sewage. The other main source of transmission is from infected food workers who have poor personal hygiene. An infected worker can transfer viral particles to any food. Therefore, proper handwashing and using a clean water supply are vital to controlling the spread of foodborne viruses.

Hepatitis A is a virus commonly associated with foodborne infections. The incubation period for hepatitis A, before a person develops any symptoms, is anywhere from 10 to 50 days. It is during this period before symptoms appear that a carrier is most infectious and most likely to spread the disease. Hepatitis A, and many other viral and bacterial pathogens, is most often transmitted via a fecal-oral route. The fact that a person is infectious even before they know they have the disease makes it difficult to control.

Parasites

Some parasites also cause foodborne illness. Parasites must live on or inside a living host to survive. The most common foodborne parasites are *Anisakis simplex*, *Cryptosporidium parvum*, *Toxoplasma gondii*, *Giardia lamblia*, and *Cyclospora cayetanensis*. *Giardia*, *Cryptosporidium*, and *Toxoplasma* are all protozoa, or single-celled organisms.

1.2.2 Chemical Hazards

Chemical hazards in food processing can include chemicals which are intentionally added to foods, incidental or unintentionally added chemicals, as well as naturally occurring toxins. Intentionally added chemicals can be preservatives, such as sulfiting agents, nutritional additives, such as niacin and color additives. Unintentionally added chemical hazards can include drug residues, unapproved food and color additives and even cleaning compounds and sanitizers commonly used in the processing facility. Naturally occurring chemical hazards include mycotoxins, such as aflatoxin in nut products; shellfish and seafood toxins; and food allergens. Control strategies for chemical hazards include effective, facility-specific Good Manufacturing Practices (cGMP's), food security and other prerequisite programs. proper labeling and understanding of all components of ingredients and rigorous control of non-ingredient chemicals.

1.2.3 Physical Hazards

Physical hazards are foreign objects such as insects, dirt, jewelry, and pieces of metal, wood, plastic, glass, etc. that inadvertently get into a food and could cause harm to someone eating that food. FDA has established maximum levels of natural or unavoidable defects in foods for substances that present no major human health hazard. These are called Food Defect Action Levels. This is the maximum amount of unavoidable defects that might be expected to be in food when handled under good manufacturing and sanitation practices. They are allowed because it is economically impractical, and sometimes impossible, to grow, harvest, or process raw products that are totally free of natural defects. Unavoidable defects include insect fragments, larvae, and eggs; animal hair and excreta; mold, mildew, and rot; shells, stems, and pits; sand and grit. The allowable levels of these substances are set at very specific levels deemed not to be a threat to human health. If a food contains more than these allowable levels, it is considered adulterated. While it may be unpleasant to find such substances in food, eating them at such low levels is not a health hazard and will not lead to illness.

1.3 History of Food Safety

Very little about foodborne illness or food safety is found in historical records. Scientists did not begin to understand bacteria, and their relationship to disease, until the late nineteenth century. People did recognize that food spoils, but the reasons for that and the potential for becoming ill from food were not known. Perhaps the absence of food safety from historical chronicles is an indication that it was less of a concern than were other problems in the past. Even early food regulations were not aimed at making food safer, but rather at preventing economic fraud. So, a history of food safety really does not exist, but numerous discoveries, inventions, and regulations have led to the present knowledge and state of affairs in food safety.

Food preservation methods such as drying, smoking, freezing, marinating, salting, and pickling had their beginnings thousands of years ago. Whether these methods were employed solely to keep food for later use, to improve flavor, or for other reasons is not known. But they also had the effect of keeping food safer. Even cooking can be viewed as an ancient method of making food safer. The Chinese Confucian Analects of 500 B. C. E. warned against consumption of sour rice, spoiled fish or flesh, food kept too long or insufficiently cooked food. The Chinese disliked eating uncooked food believing, "Anything boiled or cooked cannot be poisonous." Among the earliest of food safety manuals was one published in China in the year 2. It is possible that the practice of drinking tea originated because tea required using hot water, which would make it safer than using unheated contaminated water. Doubtless other cultures in antiquity, while oblivious to the causes or prevention of foodborne disease, experienced it and prescribed methods to avoid it.

Early scientists grappled with the nature of disease and bacteria, which would set the stage for later discoveries. Much of the present knowledge about pathogens that cause foodborne illness is built on a foundation of scientific discoveries spanning back over three centuries. Aristotle (384 – 322 B. C. E.) and his Greek philosopher/scientist predecessors believed in the spontaneous generation of organisms—that insects and animals arose spontaneously from soil, plants, or other species of animals. Francisco Redi, an Italian physician and poet, set out to disprove this theory in 1668. He believed that maggots did not arise spontaneously in meat, which challenged the common wisdom of the day. He prepared eight flasks with meat in them; four sealed and four left open to the air. No flies could land on the meat in the sealed flasks, thus no maggots grew. The clear conclusion was that maggots did not form by spontaneous generation, but that flies laid eggs that were too small to be seen. This, however, was not enough to convince skeptics. Italian biologist Lazzaro Spallanzani in 1768 disproved the spontaneous generation theory. Even though Redi proved that insects did not arise from spontaneous generation, scientists still believed that microorganisms did. In his experiments, Spallanzani boiled solutions that would normally breed microorganisms for prolonged periods of time, which killed any microorganisms that might be in the solution, on the walls of the flask, or in the air inside the flask. Then he sealed the flasks to prevent any new spores or microorganisms from entering. No microorganisms grew no matter how long he left them standing. The fact that no new microorganisms appeared meant that there was no spontaneous generation.

The discovery of bacteria in the late nineteenth century, the increased understanding of bacteria's role in disease, and the realization that there is a connection between human diseases and animal diseases led to the ideas that cleanliness is important and that unsanitary conditions can contribute to disease. In 1847 Hungarian physician Ignaz Semmelweiss wondered why women who bore their children in hospitals died of fever during childbirth, while those who gave birth at home usually did not. Noting that doctors went straight from the operating room to laboring mothers, he concluded that the doctors themselves were carrying disease to the women from the dissecting room. In those days the doctors didn't wash their hands, but wiped them on their aprons, which were already coated with body fluids. Semmelweiss ran experiments in which he had the doctors wash their hands with soap and water, and then rinse them in a chlorinated lime solution before entering the maternity wards. Death rates plummeted from 10 percent to 1.5 percent, only to climb again when the experiments were discontinued. Thereafter, he forced doctors to wash their hands before treating patients. Unfortunately, the validity of his work was not recognized at the time. His colleagues greeted his theory with ridicule, refusing to believe that their own hands were a vehicle for disease. Instead they attributed the deaths to a phenomenon arising from the "combustible" nature of the pregnant women. Historians attribute Semmelweiss's eventual despondency to the ridicule of his theories and attacks on his character. He was committed to an insane asylum, where he died of blood poisoning. Lack of personal hygiene remains one of the main causes of foodborne illness 150 years later.

In a classic case of epidemiologic sleuthing, Dr. John Snow demonstrated in 1848 how cholera spread throughout London. He noticed that people who obtained their water from a particular well were more likely to become ill than those drawing their water from another well. He persuaded city officials to remove the pump handle from that particular well, which forced inhabitants to draw water from another well. The number of cholera cases dropped immediately. Louis Pasteur further elucidated the linkage among spoilage, disease, and microorganisms with his work on fermentation and pasteurization in the 1860s and 1870s. In 1872 German scientist Ferdinand Julius Cohn published a three-volume treatise on bacteria, and essentially founded the science of bacteriology. He was the first to attempt to classify bacteria into genera and species, and the first to describe bacterial spores. But this new field of bacteriology needed bacteria on which to conduct experiments and to study. It took Robert Koch in the 1880s to perfect the process of growing pure strains of bacteria in the laboratory. At first he used flat glass slides to grow the bacteria. His assistant, Julius Richard Petri, suggested using shallow glass dishes with covers, now commonly called Petri dishes. Koch also established strict criteria for showing that a specific microbe causes a specific disease. These are now known as Koch's Postulates. Using these criteria scientists can identify bacteria that cause a number of diseases, including foodborne diseases. In 1947 Joshua Lederberg and Edward Lawrie Tatum discovered that bacteria reproduce sexually, and opened up a whole new field of bacterial genetics.

Even though Anthony van Leeuwenhook, a Dutch biologist and microscopist, had improved the microscope to the degree that small microscopic organisms could be seen for the first time as far back as 1673, the discovery of foodborne disease-causing microorganisms developed slowly. In 1835 James Paget and Richard Owen described the parasite *Trichinella spiralis* for the first time. German pathologists Friedrich Albert von Zenker and Rudolph Virchow were the first to note the clinical symptoms of trichinosis in 1860. However, the association between trichinosis and the parasite *Trichinella spiralis* was not realized until much later. In 1855 the non-pathogenic form of *Escherichia coli* was discovered. It later became a major research tool for biotechnology. Englishman William Taylor showed in 1857 that milk can transmit typhoid fever. In 1885 USDA veterinarian Daniel Salmon described a microorganism that caused gastroenteritis with fever when ingested in contaminated food. The bacteria were eventually named *Salmonellae*. August Gärtner, a German scientist, was the first to isolate *Bacillus enteritidis* from a case of food poisoning in 1888. The case was the result of a cow with diarrhea slaughtered for meat; 57 people who ate the meat become ill. Emilie Pierre-Mare van Ermengem, a Belgian bacteriologist, was the first to isolate the bacterium that causes botulism, *Clostridium botulinum*, in 1895. The case concerned an uncooked, salted ham served at a wake in Belgium. Twenty-three people became ill; three died. Van Ermengem isolated *C. botulinum* from both the ham and one of the victim's intestines. He demonstrated that the organism grows in an oxygen-free environment, and that it produces a toxin that causes the illness. In a perhaps overzealous use of the scientific method, M. A. Barber demonstrated that *Staphylococcus aureus* causes food poisoning. He became ill after each of three visits to a farm in the Philippines in 1914. Suspecting cream from a cow with an udder infection, Barber took home two bottles of cream, let them sit out for five hours, drank some of the cream, and became ill two hours later with the same symptoms he experienced on the farm. He isolated a bacterium from the milk, placed it in a germ-free container of milk, waited a while, and then convinced two hapless volunteers to drink the milk with him. They all became ill with the same symptoms (Asimov 1972). In 1945 *Clostridium perfringens* was first recognized as a cause of foodborne illness. It wasn't until the years between 1975 and 1985 that some of today's major foodborne pathogens—*Campylobacter jejuni*, *Yersinia enterocolitica*, *Escherichia coli* O157:H7, and *Vibrio cholerae*—were first recognized.

New Words

nutrient　营养物
chronic　慢性的，延续很长的
toxin　【生化】毒素
pathogen　【微生物】病菌，病原体
multidimensional　多面的，多维的
multifaceted　多面的，多维的
sanitize　清洁
bacteria　细菌
parasites　寄生的
virusin　病毒菌素

microscope　显微镜
gastrointestinal　【解】胃与肠的
infection　【医】传染，传染病，影响，感染
intoxication　中毒
multiply　繁殖，乘，增加
excrete　排泄，分泌
fecal　排泄物的，渣滓的，糟粕的
sewage　下水道，污水；用污水灌溉，装下水道于
Hepatitis　【医】肝炎
incubation　孵蛋，抱蛋，熟虑
viral　滤过性毒菌的，滤过性毒菌引起的
Anisakis simplex　简单异尖线虫
Cryptosporidium parvum　小隐孢子虫
Toxoplasma gondii　刚地弓形虫
Giardia lamblia　蓝氏贾第鞭毛虫
Cyclospora cayetanensis　圆孢子球虫
Giardia　贾第鞭毛虫
Cryptosporidium　隐孢子虫
Toxoplasma　弓形虫
protozoa　原生动物
preservative　防腐剂
niacin　烟酸的商品名
mycotoxin　【药】毒枝菌素
aflatoxin　【生化】黄曲霉毒素
allergen　【医】变态反应原，过敏原
Good Manufacturing Practices　良好操作规范
Food Defect Action Levels　食品缺陷行动水平
larval　幼虫的，幼虫状态的
mold　模子，铸型
mildew　霉，霉菌，（植物的）霉病；发霉，生霉
Chronicles　（旧约圣经）历代记
fraud　欺骗，欺诈行为，诡计，骗子，假货
marinating　腌制
plummet　垂直落下
combustible　易燃的
pregnant　怀孕的，重要的，富有意义的，孕育的
sleuth　足迹，警犬，侦探
cholera　【医】霍乱
fermentation　发酵

pasteurization 加热杀菌法，巴斯德杀菌法
treatise 论文，论述
Trichinella spiralis 旋毛形线虫
Clostridium perfringens 产气荚膜梭菌
Campylobacter jejuni 空肠弯曲菌
Yersinia enterocolitica 小肠结肠炎耶尔森氏菌
Vibrio cholerae 霍乱弧菌

Exercises

Ⅰ. Answer the following questions according to the article.

1. What is the definition of food safety? Tell us what you comprehend for food safety.
2. What is food hazard? How many hazards associated foods?
3. What types of the foodborne illness caused by pathogenic bacteria?
4. Please tell us the distinctness between infection and intoxication.
5. Please write out several parasites on the basis of your life practice.
6. What is the category of chemical hazards?
7. Write out several physical hazards from what you had seen.
8. Please talk about the safety situation in our country.

Ⅱ. Choose a term from what we have learnt to fill in each of the following blanks. Change the word form where necessary.

1. Food scontain _____ that helps prevent long-term _____ disease.
2. Consumers are the end users in the food supply chain, who are _____.
3. Food safety isa complex, _____ concept.
4. Hazards associated foods include _____, _____ and _____.
5. Biological hazards include disease-causing _____, _____ and _____.
6. Viruses differ from bacteria in that they are _____, require a living animal or human host to grow and reproduce, do not _____in foods, and are not complete cells.
7. Hepatitis A is a virus commonly associated with _____.
8. Food additives are _____ hazards, stone and iron are _____ hazards, Hepatitis A is _____ hazards.

Unit 2　Issues in Food Safety

2.1　Acrylamide

News reports of high levels of acrylamide in many fried and baked foods. These news reports have raised concerns among consumers and professionals because acrylamide is known to be toxic and can cause cancer in animals.

Acrylamide is a versatile organic compound that finds its way into many products in our everyday life. Acrylamide exists in two forms: a monomer (single unit) and a polymer (multiple units joined together by chemical bonds). The single unit form of acrylamide is toxic to the nervous system, a carcinogen in laboratory animals, and a suspected carcinogen in humans. The multiple unit or polymeric form is not known to be toxic.

Traditionally, human exposure to acrylamide (monomeric form) was believed to occur only in workplaces or other environments with little relevance to the public at large. However, acrylamide can be absorbed through skin contact, breathed in, or consumed in contaminated foods or water. Both the World Health Organization (WHO) and the US Environmental Protection Agency (EPA) have set the maximum contaminant levels for acrylamide in drinking water at 0.5 parts per billion (ppb) or at 0.5 microgram per liter. Smokers absorb the chemical from tobacco smoke, as can those exposed to second hand smoke. Currently, data on acrylamide intake from foods and its fate in the body after the intake are limited.

Once inside the body, acrylamide binds to red blood cells. Potential symptoms of overexposure to monomeric acrylamide include numbness of the limbs, and weakness and lack of coordination in the legs. Long-term exposure to small doses of acrylamide causes nerve damage in the extremities. Some tunnel construction workers have experienced neurological damage from exposure to acrylamide in grout. Animal studies have shown acrylamide to be a carcinogen although cancer in humans following occupational exposures has not been reported.

2.2 Food Allergies

A food allergy is an acquired hypersensitivity reaction to what is normally considered a safe food. Food allergies occur more often in children than in adults: 5% ~ 8% of those age 4 or under and about 1% ~ 4% of adults are affected. Those with severe reactions may experience what is known as anaphylaxis or *anaphylactic shock*. Annually, around 30,000 people receive life-saving emergency treatment and 150 fatalities occur.

While most food allergies in adults are caused by a small group of foods or food products, early in life food allergies can be caused by a wider variety of foods. About 90% of reported food allergies in children under the age of four are caused by: dairy products, tree nuts, eggs, wheat and wheat products, peanuts, or soy and soy products.

Dairy, eggs, and soy allergies are commonly outgrown; peanut allergies are almost never outgrown. As an adult, "the big eight" foods (and their products) account for 90% of food allergies: cereals containing gluten, crustaceans, milk, eggs, tree nuts, fish, soybeans and peanuts.

A true allergy is caused by a person's immune system reacting to a food when first eaten. The body "remembers" that food and, when it is eaten again, the immune system overreacts in an excessive and potentially life-threatening way.

Allergies can also be classic, also known as "atopic". Instead of an individual developing an allergy spontaneously, they inherit a predisposition to develop food or other allergies. Often, these individuals suffer from hay fever (allergic rhinitis), asthma, or rashes (atopic dermatitis) and are more likely to develop a food allergy.

True food allergies may lead to severe allergic reactions or anaphylactic shock caused by rapid release of Immunoglobulin E (IgE). IgE is a natural component of the immune system, normally involved in protecting the body from parasites. However, when over-production of IgE is triggered by a food or other allergen (any material that triggers an allergic response), local or systemic (effecting the whole body) inflammation, severe swelling, or hypersensitivity reactions may occur.

Allergic reactions to food may cause symptoms within seconds of consumption, or the symptoms may take up to several hours to develop. Symptoms can occur locally, or can be spread over the body or in multiple locations. Redness, itching, and swelling (inflammation) are the most well known and commonly associated symptoms, although several other types of symptoms are possible.

Symptoms associated with the digestive tract may include any one or more of the following:
- itching/tingling of the lips, palate, tongue, or throat;
- hoarseness and sensation of tightness in throat;
- vocal impairment or difficulty speaking;
- swelling of the lips or tongue;
- abdominal pain or cramps;
- nausea and/or vomiting; or
- diarrhea.

If a person has an allergy to a particular food, any meal with that food present, even as a flavoring, may cause an allergic response. If a person is allergic to peanuts, they will be sensitive to the consumption of any food that has peanuts or peanut products (peanut butter, peanut oil, chopped or diced peanuts, etc.) as an ingredient. Treatment or processing of a food does not affect its ability to cause an allergic response. **It is important to carefully read food labels and ingredient lists if a person has a known food allergy.**

There is no cure for a food allergy. Once diagnosed, a person will most likely have to contend with their condition for life. **If a person has an allergy to a particular food, the only proven therapy is strict avoidance of the food or its products.**

If an attack occurs, the medication of choice is an injection of epinephrine. A person with a known food allergy should always carry a dose of epinephrine in case of an emergency.

A person suffering from an anaphylactic attack should be taken by an ambulance to a hospital. Even if epinephrine is administered to the victim, anaphylactic symptoms may reappear within minutes, or several hours, after treatment. Observation by trained medical personnel is important during this period.

2.3 Genetically Modified Food

A genetically modified (GM) food is a result of recombinant DNA biotechnological procedures that allow the genetic makeup of an organism to be modified. This can be accomplished by incorporating genes from other organisms or by rearranging genes already present. These changes can result in the expression of attributes not found in the original organism. Examples of products that have been engineered include delayed-ripening tomatoes; pest-resistant crops, such as virus-resistant squash and Colorado potato beetle-resistant potato; herbicide-tolerant crops, such as bromoxynil-tolerant cotton and lyphosate-tolerant soybean; and many others.

Genetically modified (GM) organism is one that has had its genetic material altered through any method. A **genetically engineered** (GE) organism is one that is modified using techniques that permit the direct transfer or removal of genes in that organism. Such techniques are also called recombinant DNA or rDNA techniques. Lastly, **transgenic** organisms have a gene from another organism moved into them.

This practice has raised ethical issues as well as concerns about possible health implications. Groups opposed to the genetic manipulation of food have termed this practice and its subsequent products "Frankenfood". Though more research is needed, the FDA feels that there are no serious food safety issues associated with these products, although they are always alert for possible food allergens.

The potential for GM foods to cause allergic reactions is the most obvious health concern associated with these products. Specific proteins in milk, eggs, wheat, fish, tree nuts, peanuts, soybeans, and shellfish cause over 90% of food allergies. If a protein from one of these food types were to be incorporated into a food that normally would not have this protein, people who are allergic to these proteins could unknowingly consume such a food and suffer allergic reactions. The FDA has put measures into place to prevent such a scenario by requiring that each producer of a GM food product present scientific evidence that they have not incorporated any allergenic substance into their product. If this evidence cannot be produced, the FDA requires a label to be put on the product to alert consumers.

With all the controversy surrounding GM foods, especially in Europe, researchers have been searching for new methods to enhance crop production. The newest technique is called marker-assisted selection (MAS). This product combines traditional genetics and molecular biology. MAS allows for the selection of genes that control traits of interest, such as color, meat quality, or disease resistance. It has the promise of becoming a valuable tool in selecting organisms for these traits of interest. Because this process uses existing DNA, not transgenic DNA, to choose desired traits, MAS stands to be less controversial then other GM techniques.

2.4 Mad Cow Disease

Mad Cow Disease is considered to be old news; however the problem in the United Kingdom is not over yet. It has caused some very severe problems for farmers in the UK. It has also resulted in some trade relations problems between the UK and the rest of the world.

It is a disease of the brain that causes cattle to slowly go crazy. The cattle become very unstable and can't stand up, act unpredictable, and eventually, they die. The disease basically destroys the brain It is officially called Bovine Spongiform Encephalopathy (BSE). Encephalopathy means that damage is occuring in the brain itself. Spongiform is the term used to describe what happens to the cow's brain. When a cow's brain is removed at autopsy the laboratory finds the diseased parts of the brain to be **spongy** in appearance. The first well-documented case of BSE was reported in 1986.

Mad Cow Disease or BSE is caused by a **prion** (**pro**teinaceous **in**fectious particle). This term is an acronym coined by Dr. Stanley Prusiner. This disease and others like it are transmitted by a protein. These diseases are very unusual. Viruses, bacteria, fungi, or parasites do not cause these diseases. Humans can also get Spongiform Encephalopathies (ex. kuru, Creutzfeldt-Jakob disease (CJD), and Gerstmann-Straussler-Scheinker syndrome).

This prion now has a name: **PrP**. PrP is found in a normal form in all animals. An altered form of PrP (PrP-sc) is what causes BSE. Cattle feed containing PrP-sc from sheep is believed to be the source of the infection in the cattle. To boost the protein content of cattle feed, manufacturers of feed placed sheep parts in the feed. Sheep also can get a spongiform encephalopathy called **Scrapie**. It is currently believed that somehow PrP-sc changed enough to cause infection in cattle. The use of sheep in cattle feed was banned in 1989. However, farmers continued to feed cattle the sheep-containing feed well into the 1990s.

There is no clear-cut answer to that question. Some current experiments have shown that BSE-infected beef when fed to sheep can cause BSE. A number of carnivores in various British zoos have also developed BSE when fed presumably BSE-contaminated beef. In other words BSE can cause infection in other animals besides cattle.

No one knows for sure if humans can get BSE from this beef. However, in March of 1996, ten cases of a different form of Creutzfeldt-Jakob disease (CJD) were reported in humans inBritain. The average age of these patients was 28 and the patients did not have the same symptoms normally seen in CJD patients. The average age of people with CJD is normally about 63. When the laboratories looked at the brains of the patients with the new type of CJD the brains looked more like the pattern seen with BSE in cows than classic CJD. An association? Yes. Solid proof? No! However, to be on the safe side many countries no longer import British beef.

2.5　Food Irradiation

Food irradiation is a method of controlling insect pests and pathogenic or spoilage bacteria in food and agricultural commodities. Instead of using heat or chemicals for processing, irradiation uses gamma energy, electron beams, or X-rays. Sometimes it is referred to as cold pasteurization. Similar technology is used to sterilize medical equipment and devices so they can be used in surgery and implanted without risk of infection.

Food irradiation has different uses depending on the strength of radiation used. It can be used to control mold, inhibit sprouting in vegetables, control insect pests, reduce bacterial pathogens, or, at the strongest dose, sterilize food. At low doses irradiation is an alternative to fumigation with chemicals to eliminate insects. Low doses have been used to inhibit the growth of mold in strawberries and to inhibit sprouting in potatoes, thereby prolonging the shelf life of these products. The current push for irradiation is to kill bacteria and parasites that would otherwise cause foodborne illness. The dose of irradiation needed to kill *Salmonella* in chicken is about seven million times more than that of a chest X-ray. NASA irradiates food that astronauts eat in space. Their food is irradiated to the level of sterilization.

Three different irradiation technologies exist, which use three different kinds of rays: gamma rays, electron beams, and X-rays. The first technology uses the radiation given off by a radioactive substance. This can be either a radioactive form of the element cobalt (cobalt 60) or of the element cesium (cesium 137). These substances give off high-energy photons, called gamma rays, which can penetrate foods to a depth of several feet. These particular substances do not make anything around them radioactive. This technology has been used routinely for more than 30 years to sterilize medical, dental, and household products. It is also used for radiation treatment of cancer.

Electron beams, or e beams, are produced in a different way. The e beam is a stream of high-energy electrons, propelled out of an electron gun. This electron gun apparatus is a larger version of the device in the back of a TV tube that propels electrons into the TV screen at the front of the tube, making it light up. The electron beam generator can be simply switched on or off. No radioactivity is involved. The electrons only penetrate food to a depth of a little more than an inch, so the food to be treated must be no thicker than that. Two opposing beams can treat food that is twice as thick. E-beam medical sterilizers have been in use for at least 15 years.

The newest technology is X-ray irradiation. This is an outgrowth of e-beam technology, and is still under development. The X-ray machine is a more powerful version of the machines used in many hospitals and dental offices to take X-rays. To produce the X-rays, a beam of electrons is directed at a thin plate of gold or other metal, which produces a stream of X-rays coming out the other side. Like gamma rays, X-rays can pass through thick objects. However, like e beams, the machine can be switched on and off, and no radioactive substances are involved.

Irradiation kills microbes by damaging their DNA. Because of this, bigger organisms like parasites and insects are more susceptible to irradiation because they have more DNA. A higher dose is necessary to kill bacteria since they have less DNA. Viruses, which are very small and have very little DNA, are generally resistant to irradiation at the doses approved for use in foods. Not all foods are suitable for irradiation. The quality of some foods, such as eggs and shellfish, decreases below the point of consumer acceptability. Irradiation also changes some of the taste and texture qualities of foods. The higher the dose, the more pronounced the changes, just like in conventional cooking. The technology to irradiate foods to make them safer has been available for decades, but it languished in legislative limbo brought about by those opposed to it.

The current flurry of activity in favor of food irradiation started after the 1994 deaths of four children who died after eating hamburger meat contaminated with *E. coli* O157:H7 from a fast food restaurant. The 1997 recall of 25 million pounds of hamburger meat from a major meat processing plant in Nebraska added fuel to the fire. Proponents of food irradiation claim that the safety of irradiated foods has been studied by feeding them to animals and to people. These extensive studies include animal feeding studies lasting for several generations in several different species, including mice, rats, and dogs. There is no evidence of adverse health effects in these well-controlled trials. Irradiating foods does produce a very small amount of unique radiation products—about three milligrams per kilogram of food, equivalent to three drops in a swimming pool.

Irradiation proponents emphasize that it is not a substitute for good sanitation. For irradiation to be effective, the food that is to be irradiated already needs to be clean. The more initial contamination there is, the higher dose of irradiation it would take to eliminate possible pathogens, and the greater the change in the taste and quality of the food. So, irradiating poor-quality or spoiled food will result in a product that will be of poorer quality after irradiation, making it impossible to sell. It is in the best interest of industry to irradiate clean, good-quality food. Irradiation adds an extra measure of protection to food, and is only one tool in the arsenal to fight foodborne pathogens.

2.6 Pesticides

Pesticides are designed to control pests, but they can also be toxic (poisonous) to desirable plants and animals, including humans. Some pesticides are so highly toxic that very small quantities can kill a person, while exposure to a sufficient amount of almost any pesticide can make a person ill. Since even fairly safe pesticides can irritate the skin, eyes, nose, or mouth, it is a good idea to understand how pesticides can be toxic so you can follow practices designed to reduce or eliminate your exposure and the exposure of others to them.

Before a pesticide can harm you it must be taken into the body. Pesticides can enter the body orally (through the mouth and digestive system); dermally (through the skin) or by inhalation (through the nose and respiratory system).

Toxicity refers to the ability of a poison to produce adverse effects. These adverse effects may range from slight symptoms such as headaches to severe symptoms like coma, convulsions, or death. Poisons work by altering normal body functions. Most toxic effects are naturally reversible and do not cause permanent damage if prompt medical treatment is sought. Some poisons, however, cause irreversible (permanent) damage.

All new pesticides are tested to establish the type of toxicity and the dose necessary to produce a measurable toxic reaction. In order to compare the results of toxicity tests done in different labs, there are strict testing procedures. Toxicity testing is extensive (involving many phases) and therefore expensive. Humans, obviously, cannot be used as test subjects, so toxicity testing is done with animals and plants. Since different species of animals respond differently to chemicals, a new chemical is generally tested in mice, rats, rabbits, and dogs. The results of these toxicity tests are used to predict the safety of the new chemical to humans.

Toxicity tests are based on two premises. The first premise is that information about toxicity in animals can be used to predict toxicity in humans. Years of experience have shown that toxicity data obtained from a number of animal species can be useful in predicting human toxicity, while data obtained from a single species may be inaccurate. The second premise is that by exposing animals to large doses of a chemical for short periods of time, we can predict human toxicity from exposure to small doses for long periods of time. Both premises have been questioned.

Toxicity is usually divided into two types, acute or chronic, based on the number of exposures to a poison and the time it takes for toxic symptoms to develop. Acute toxicity is due to short-term exposure and happens within a relatively short period of time, whereas chronic exposure is due to repeated or long-term exposure and happens over a longer period.

The commonly used term to describe acute toxicity is LD_{50}. LD means *lethal dose* (deadly amount) and the subscript 50 means that the dose was acutely lethal to 50% of the animals to whom the chemical was administered under controlled laboratory conditions. The test animals are given specific amounts of the chemical in either one oral dose or by a single injection, and are then observed for a specified time.

The lower the LD_{50} value, the more acutely toxic the pesticide. Therefore, a pesticide with an oral LD_{50} of 500 mg/kg would be much less toxic than a pesticide with an LD_{50} of 5 mg/kg. LD_{50} values are expressed as milligrams per kilogram (mg/kg) which means milligrams of chemical per kilogram of body weight of the animal. *Milligram* (mg) and *kilogram* (kg) are metric units of weight. Milligrams per kilogram is the same as parts per million. To put these units into perspective, 1 ppm is analogous to 1 inch in 16 miles or 1 minute in 2 years.

LC means *lethal concentration*. Concentration is used instead of dose because the amount of pesticide inhaled in the air is being measured. LC_{50} values are measured in milligrams per liter. Liters are metric units of volume similar to a quart. The lower the LC_{50} value, the more poisonous the pesticide.

Chronic toxicity refers to harmful effects produced by long-term exposure to pesticides. Less is known about the chronic toxicity of pesticides than is known about their acute toxicity, not because it is of less importance, but because chronic toxicity is much more complex and subtle in how it presents itself. While situations resulting in acute exposure (a single large exposure) do occur, they are nearly always the result of an accident or careless handling. On the other hand, persons may be routinely exposed to pesticides while mixing, loading, and applying pesticides or by working in fields after pesticides have been applied.

There is no standard measure like the LD_{50} for chronic toxicity. How chronic toxicity of chemicals is studied depends upon the adverse effect being studied. Chronic adverse effects may include carcinogenesis, teratogenesis, mutagenesis, blood disorders (hemotoxic effects), endocrine disruption, and reproductive toxicity.

Carcinogenesis means the production of malignant tumors. Oncogenesis is a generic term meaning the production of tumors which may or may not be carcinogenic. The terms tumor, cancer, or neoplasm are all used to mean an uncontrolled progressive growth of cells. In medical terminology, a cancer is considered a malignant (potentially lethal) neoplasm. Carcinogenic or oncogenic substances are substances which can cause the production of tumors. Examples are asbestos and cigarette smoke.

Teratogenesis is the production of birth defects. A teratogen is anything that is capable of producing changes in the structure or function of the offspring when the embryo or fetus is exposed before birth. An example of a chemical teratogen is the drug thalidomide which caused birth defects in children when their mothers used it during their pregnancy. Measles virus infection during pregnancy has teratogenic effects.

Mutagenesis is the production of changes in genetic structure. A mutagen is a substance which causes a genetic change. Many mutagenic substances are oncogenic, meaning they also produce tumors. Many oncogenic substances are also mutagens.

New Words

acrylamide   【化】丙烯酰胺
toxic   有毒的，中毒的
versatile   通用的，万能的
polymer   聚合体
carcinogen   致癌物质
allergy   【医】敏感症，过敏症
atopic   【医】遗传性过敏症的
rhinitis   【医】鼻炎，鼻粘膜炎
asthma   【医】哮喘
dermatitis   【医】皮炎
anaphylactic   过敏的，导致过敏的
immunoglobulin   【生化】免疫血球素，免疫球蛋白
inflammation   【医】炎症，发炎
hypersensitivity   超敏性
itching   痒的
swelling   发炎
abdominal   腹部的
cramp   抽筋，腹部绞痛，月经痛
nausea   反胃，晕船，恶心，作呕，极度的不快
diarrhea   痢疾，腹泻
epinephrine   【生化】肾上腺素
genetically modified (GM) food   转基因食品
Mad Cow Disease   疯牛病
spongy   像海绵的，柔软，多孔而有弹性的
prion = proteinaceous infectious particle
kuru   【医】苦鲁病，新几内亚震颤病
scrapie   （羊患的）痒病
food irradiation   食品辐照
cobalt   【化】钴（符号为Co）
cesium   【化】铯（符号为Cs）
pesticide   杀虫剂
dermal   皮肤的，真皮的
oncogenesis   【医】瘤形成
malignant   恶性的

neoplasm 【医】赘生物,(肿)瘤
tumor 瘤
asbestos 【矿】石棉
teratogenesis 畸形生长
teratogen 【生】致畸剂,畸胎剂
embryo 胚胎,胎儿,胚芽;胚胎的,初期的
fetus 胎儿
thalidomide 镇静剂,安眠药之一种
measles 【医】麻疹,风疹,包虫病,痧子
teratogenic 【生】产生畸形的
mutagenesis 【生】突变形成,变异发生

Exercises

Ⅰ. Answer the following questions according to the article.

1. How many forms have a crylamide? What are they?
2. Please talk about the toxicity of a crylamide.
3. What is food allergy? Which foods have allergy? Please list eight food allergies.
4. What are the symptoms caused by allergic reactions to food?
5. What is genetically modified (GM) food?
6. What is the reason that causes Mad Cow Disease?
7. What is food irradiation? There are three different kinds of rays, what are they?
8. How to do toxicity tests of pesticides?

Ⅱ. Choose a term from what we have learnt to fill in each of the following blanks. Change the word form where necessary.

1. The _____ unit or _____ form of acrylamide is NOT known to be toxic.
2. Exposing to acrylamide in grout could cause _____ damage.
3. Food allergies occur more often in _____ than in _____.
4. Food allergy may lead to _____ by rapid release of Immunoglobulin E.
5. About 90% of reported food allergies in children under the age of four are caused by: _____
_____.
6. _____ are the most well known and commonly associated symptoms.
7. A _____ organism is one that is modified using techniques that permit the direct transfer or removal of genes in that organism.
8. Mad Cow Disease or BSE is caused by a _____.
9. Irradiation uses _____, _____, _____ for processing.
10. The lower the _____ value, the more acutely toxic the pesticide.

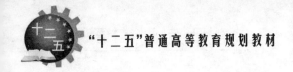

Unit 3 GMP and SSOP

Good Manufacturing Practices (GMPs) are federal law. GMP tell us what a processing establishment shall do in terms of buildings, facilities, equipment, production and process controls, warehousing, and distribution.

Sanitation Standard Operating Procedures (SSOPs) specify how the processor will meet the sanitation conditions and practices that are to be monitored. Each processor shall monitor the conditions and practices during processing with sufficient frequency to ensure, at a minimum, conformance with those conditions and practices specified in that chapter to be appropriate to the plant and the food being processed.

3.1 Good Manufacturing Practice

Good Manufacturing Practice Regulations were implemented by the Food and Drug Administration (FDA), to establish manufacturing standards for food products. These regulations were mandated to ensure the safety and wholesomeness of the processed food supply. All areas of food processing are covered in this publication to include personnel, buildings and facilities, equipment, and production and process controls. Each of these areas will be discussed.

3.1.1 Personnel

The food plant personnel are vital to a company's success. It is management's responsibility to educate personnel on the fundamental principles of food plant sanitation, and the importance of personal hygiene. Regulations state that necessary measures be taken to ensure disease control, cleanliness, and proper education, training, and supervision to enforce compliance. The following are strategies to help in personnel education.

Personal Cleanliness
- Daily bathing is required before work.
- Hair must be washed at least once a week.
- Nails should be kept clean and properly trimmed.
- Jewelry is not permitted on the production floor.
- Disposable gloves must be worn if bandages or cuts are present on hands.
- Communicable illnesses must be reported; personnel affected or suffering from open lesions or infected wounds should not be allowed to work with edible food products.

Uniforms and Underclothing
- Uniforms/frocks must be kept clean and neat.
- All frocks and equipment will be removed before using restrooms.
- Frocks will not be worn outside of the plant.
- Frocks worn in raw product area must be removed; and clean frocks worn in cooked product areas.

- Pockets are not permitted above the waistline.
- Protective shoes and eyewear should be worn if appropriate.
- Sweaters (or like clothing) should be avoided or covered by a uniform.
- Uniforms should be changed if soiled.
- Pants must be tucked into boots.
- Boots need to be cleaned before entering processing area.

Hair Restraints

- Hair must be covered; hair bonnets are preferred.
- Hair bonnets should be new and unused. Each time a bonnet is removed it should be discarded.
- Males must be clean-shaven or facial snoods are necessary. Moustaches are permitted if trimmed and above the corners of the mouth.
- Sideburns must be covered above ear lobes.

Handwashing

- Hands must be washed following proper handwashing procedure.
- Hands must be washed after:
  —Coughing or sneezing
  —Using the restroom
  —Smoking
  —Breaks
  —Handling soiled containers, waste materials, or animal products
  —Using the telephone
- Handwashing must be made convenient. Sinks with hot water are necessary for proper washing habits.
- All personnel equipment should be cleaned at the end of each shift, or more often if necessary.
- Wall-mounted dispensers of anti-bacterial soap and sanitizing solution should be placed beside sinks, and rolls of clean disposable towels made available.
- To minimize contact with germs on faucet handles by dirty hands, workers should be trained to shut off the water with the towel after drying their hands.
- If there is a door in the handwashing area, employees should open the door with the towel, then dispose of the towel as they exit the room.

Conduct

- Running, horseplay, and riding on equipment are not allowed.
- Spitting, smoking and chewing tobacco are not allowed.
- Urinals and toilets must be flushed after each use.
- Maintenance tools or parts are not permitted on food contact surfaces.
- Eating or drinking must be done in specified areas, separate from food processing area.
- Lockers should be kept clean and orderly.

Education and Training

Employees that lack the necessary training have proven to be detrimental to a quality assurance program. Human error is frequently the cause of product failure and noncompliance. Every position has a unique set of manual, technical, and administrative skills necessary to complete the job effectively. The level of competency necessary for a task depends on the position.

For personnel handling food, it is especially important to train in the area of sanitation. This training should emphasize the importance of personal hygiene, proper food handling procedures, proper sanitation, record keeping, testing, and process procedures. Training should be done when the employee is hired and continued throughout the length of their employment. It is necessary to take the education level and prior training of the employee into consideration. Present material to the employee at the appropriate level and in a variety of methods to include: visual signs, videos, lectures by speakers, demonstrations, role-playing and hands-on-training.

3.1.2 Buildings and Facilities

Plant and Plant Grounds

The regulations for maintenance of buildings and facilities refer to the structures under the company's control. The immediate vicinity of a facility must be kept clean from trash and refuse. The roads and parking lots associated with the facility should be paved to avoid unintentional contamination. Grass or weeds around the facility should be mowed or kept short to eliminate breeding or harboring of pests. This is especially important as rodents, birds, and insects carry numerous types of disease that can be transmitted to or hazardous to humans. Adequate drainage of facility grounds is essential to eliminate seepage, tracking of dirt, and pest breeding grounds. If problems are present in areas not under the company's control, appropriate measures need be taken to ensure that those areas will not present any type of contamination (i.e. extermination of pests).

Plant Construction and Design

The plant needs to be easily cleaned and sanitized. The placement of equipment has a direct impact on ease of cleaning and accessibility. By providing sufficient room for proper cleaning and sanitizing, the process will be much easier. The floors, walls, and ceilings should be easily cleaned and kept in a sanitary condition. Floors should be slightly sloped to allow for proper drainage and avoid water accumulation. Fixtures, ducts, and pipes should be suspended away from working areas and aisles, and working areas kept unobstructed. Ventilation and lights should be adequate; and lights enclosed in safety fixtures to avoid contamination due to breakage. Separating the food processing area from the rest of the facility is a necessity to reduce the potential for contamination. To minimize pests, door and windowsills should be tight. Windows should be protected with screens, as well as other openings that would allow unwanted pests. Drains need traps and proper covers or grills.

Sanitation

The most important aspect in sanitation is the commitment to producing safe, wholesome products in a clean plant environment. This commitment must come from management and be communicated to all employees. It is important to assign at least one employee to be responsible for sanitation practices in the manufacturing facility. Support of this employee's needs is essential, and management should require daily reports from this employee. This employee should have training in the areas of microbiology, chemistry, and entomology, whether it is in the form of a degree or short course on these subjects. Once a sanitation program is established, it is important that improvements be made as new developments continually arise. The sanitation program's top priority should be to communicate to all employees the necessity and importance of proper cleaning and sanitation practices. *The Code of Federal Regulations*, *Part* 110, *Subpart B*, gives specific guidelines concerning sanitary operations, facilities, and controls. These should be read and fully understood by management and sanitation personnel.

Equipment

While each processing facility has different pieces of equipment that is specific for the food it is producing, certain factors are universal when designing and installing equipment. Since this equipment must produce sanitary food products, it is important to plan and operate following specific guidelines. General guidelines are listed below and further reference should be made to the *Code of Federal Regulations*, *Part* 110, *Subpart C*.

General Guidelines

- Food contact surfaces should be inert under conditions of use, smooth and non-porous. Stainless steel is preferred, wood is not permitted. All surface seams should be smooth, continuous, and flush with surface.
- Equipment contact surfaces should be easily cleaned and sanitized by using access doors, removable covers or disassembly.
- Equipment assembly parts such as bolts, nuts, washers, and gaskets should be kept away from food during equipment operation. Moving parts should have sealed or self-lubricating bearings.
- Installation of equipment should allow for 3 feet of clearance around, and 6 inches off of the floor for the working area and proper cleaning. Equipment should be installed considering convenience, serviceability, and maintenance.
- Clean in place systems are preferred over cleaning out-of-place systems.
- Motors, pulleys and drums should be fully enclosed and sealed, and not mounted directly above food contact surfaces.
- Conveyors and conveyor parts need to be fully accessible for easy cleaning.
- Valves for water and steam should not allow for leakage; and valves for food should be easily disassembled for cleaning and inspection.
- Pipes, irons, and beams should be installed following very specific guidelines. These are referenced in the *CFR*, *Part* 110, *Subpart C*.
- Kettles or cookers require lids and a self-draining design.

3.1.3 Production and Process Controls

Each processing facility will have a unique process specifically designed for the product that it produces. The CFR guidelines should be used to reference specific needs but some general guidelines are listed below.

- All operations in receiving, transporting, packaging, preparing, processing, and storing of food must follow sanitary principles.
- Raw materials must be inspected and separated from processed products.
- Raw material containers need to be inspected.
- Ice, if used, must be sanitary if used for food production.
- Food processing equipment should be inspected and cleaned on a regular basis.
- Processing factors such as time, temperature, humidity, pressure, and other relevant variables must be properly controlled and documented.
- Testing procedures must be in place to check finished product for quality and safety.
- Packaging materials must be approved and provide appropriate protection.
- Finished products must be coded to provide information such as place and date of production.
- Production records need to be correctly maintained and retained for an appropriate amount of time specified in the CFR.
- Products must be warehoused and shipped under sanitary conditions and away from harmful substances.

By following these guidelines and those mandated in the *Code of Federal Regulations*, a company can manufacture a product that meets the needs of their consumers and operate the processing facility in a safe and efficient way. By doing this, employees will be more productive and have fewer accidents. Consumers will be less likely to complain about products, in turn benefiting both the consumer and producer.

3.2 Sanitation Standard Operating Procedures

In addition to preparing a suitable GMP program in a plant, plants are required to develop their own, individual Sanitation Standard Operating Procedures. Even if not required, these will be quite useful in conjunction with HACCP programs that may also be voluntary. SSOPs are to describe every cleaning and sanitation procedure used in a plant, in detail. The approximate time of the day or time in a shift when each operation is to be performed is specified. Water analyses and temperatures are likely to be included, as well as specific cleaning and sanitizing agents, how they are used, and on what. The person responsible for each step is to be specified, usually by job title, rather than by name. Provision is made for documenting every operation: the person who does it, subject to periodic verification by a supervisor, records the performance of each required task. As we will see, unit operations should be covered in the SSOP rather than in the HACCP plan, if any.

According to FDA, the SSOP should specify how the processor would meet those sanitation conditions and practices that are to be monitored with sufficient frequency to ensure outcomes appropriate to the plant and the food being processed. According to the National Conference on Interstate Milk Shipment (NCIMS), prerequisites contain eight topics; the main subjects in this category include:

——Safety of water

——Conditions and cleanness of food contact surfaces

——Prevention of cross contamination

——Maintenance of hand washing/sanitizing & toilet facilities

——Protection of food materials from adulteration

——Proper labeling, storage and use of toxic components

——Control of employee health condition

——Exclusion of pests

All of the above items should be addressed by the following four aspects: Control Measures, Monitoring Procedures, Corrective Actions and Record Keeping.

Examples in dairy industries:

Prerequisite 1: Safety of water monitoring

Sources of water in a plant have two main possibilities—municipal and private—either of which requires an appropriate plan based on intended usage. It is necessary to have a fundamental program for private (well water or other) sources. In EU HACCP standards, well water is designated as a CCP. Water mainly will be used as an ingredient in dairy production, to convey or transport products, to rinse the products, to make ice, in cooling water systems and steam and to clean and sanitize facilities, utensils, containers and equipment. The goal is mainly safe supply, for water that contacts food and food contact surfaces and for production of ice, cooling water systems and steam systems. For municipal water, the microbiological control could be done each month; and it will also be necessary to analyze annually heavy metals and other chemical contaminants in the water. It is strongly recommended to have a well documented system to control all of the related elements in regard to water supplements.

For ice: Monthly water microbiological test will be done by Quality Assurance (QA).

Cooling water: Micro (QA) + maintenance of tank (monthly) + state micro result (3 months).

Steam: Additive verification (annual) (QA) and correction; inspect system and verify chlorinator dosing properly.

Prerequisite 2: Condition and cleanness of food contact surfaces

Condition and construction of food contact surfaces with application and establishments of the related standards such as 3-A standards: Clean in place (CIP) systems including time, temperature, pressure, concentration of the cleaning material that will be used for sanitation; Clean out of place (COP) systems, including time, temperature and concentration, appropriate chemicals based on the regulations, which are approved by Environmental Protection Agency (EPA). Gloves and garments should be clean and easily cleanable and in good condition when in use, and for hands contacting food, application of disposable gloves, if necessary.

For monitoring these processes: daily CIP, COP time, temperature, pressure, concentration and daily pre-operational inspection and weekly microbiologic testing. For cleaning programs, daily, weekly, and monthly programs are advised. Food contact surfaces are cleaned twice each day, once at the midday break, and again at the finish of the shift. Examples for the Standard Operating Processes (SOP) are indicated below:

Pick up and remove major solid wastes. Tear down sluices.

Rinse all surfaces with cold water. Degrease and scrub all surfaces with degreaser.

Rinse all surfaces with hot water (190°F).

Sanitize with quaternary ammonia. This food contact-approved sanitizer is squeezed from surfaces prior to resuming production after the midday break. It is left on overnight at the end of shift cleaning.

Floors are freed of solid waste, then rinsed twice each day with hot water (190°F) and sprayed with quaternary sanitizer.

Utensils are cleaned twice each day, once at the midday break and again at the finish of the shift.

Clean in deep sink with degreaser (and abrasive cleaner if necessary).

Rinse in hot water (190°F). Soak in quaternary sanitizer.

Rinse in hot water (190°F) prior to use.

Prerequisite 3: Prevention of cross contamination

This is done by: suitable separation of raw and processed product and allergenic ingredients; application of color coding of brushes; control programs for employee hygiene, dress and hand washing practices; programming for employee movement related to product flow; and the proper segregation and disposal of waste products. Monitoring is based on inspecting the plant for adequate segregation of raw from processed aterials, cleaning brushes, utensils and allergenic ingredients. All finished-product workers are required to wear hair/beard nets, smocks, and disposable gloves. Raw-product workers also wear waterproof aprons.

Prerequisite 4: Maintenance of hand-washing and toilet facilities

Hand-washing stations and toilet facilities shall be maintained to prevent cross contamination of filth and potential pathogens from employees to food and food-contact surfaces. There are some opinions that use of gloves is not as effective as is generally believed. Hand-washing processes may differ; one of the suggested processes is described below:

Turn on the water.

Put 3 to 4 mL of plain liquid soap on the nail brush.

Underneath the flowing water in the sink, for about 10 to 12 seconds, lightly brush the fingertips with the tips of the brush bristles while the water flows over the brush and fingertips. The bacteria come off and go down the drain. This gives a 1,000-to-1 reduction of bacteria on fingertips.

Put the brush down. Put soap on the hands, and lather the hands underneath the flowing water for about 5 seconds to get a 100-to-1 reduction. Rinse the soap from the hands.

Paper-towel dry for another 10-to-1 reduction.

Prerequisite 5: Protection from adulteration

The goal of this system would be to ensure that the food, food-packaging material and food-contact surfaces are protected from adulteration with lubricants, fuel, pesticides, cleaning components, sanitizing agent, condensates, and other chemical, biological and physical contaminants. It is necessary to have a regular (monthly, for instance) program to control all of the above materials in all of the processing and storage areas.

Prerequisite 6: Proper labeling, storage and use of toxic components

This is intended to ensure that toxic components are properly labeled, stored and used and that these measures are adequate to protect food from contamination. It is necessary to have a regular (monthly, for instance) program to determine that all of the above materials are properly labeled, stored and used in the plant.

Prerequisite 7: Control of employee health

Ill, injured and bandaged personnel should be monitored regularly to avoid contamination of food, food-packaging materials and food-contact surfaces. This control should be done daily with a checklist and annually by means of a training course, to train personnel concerning foodborne diseases and related controls.

Prerequisite 8: Exclusion of pests

Plans and monitoring should confirm that presence/absence of pests, pest harborage, building design, openings and doors, waste disposal, pesticide use, and housekeeping are suitable to control the pests.

New Words

GMP = Good Manufacturing Practices 良好操作规范
SSOP = Sanitation Standard Operating Procedures 卫生标准操作程序
mandate （书面）命令，训令，要求，
hygiene 卫生，卫生学
trim 整齐的，整洁的；整理，修整，装饰
frock 上衣，外衣，工装
waistline 腰围，腰身部分
bonnet 无边女帽，童帽，烟囱帽，阀帽
ear lobe 耳垂
faucet 龙头，旋塞
horseplay 动手脚和大声欢笑的玩闹
urinal 尿壶，小便池
toilet 盥洗室，梳洗
locker 上锁的人，有锁的橱柜，锁扣装置，有锁的存物柜
trash 垃圾，废物 = refuse

harbor 海港,滋生地

mow 扫除,收割庄稼,扫倒

seepage 渗流,渗出的量

duct 管,输送管,排泄管

unobstructed 不被阻塞的,没有障碍的,畅通无阻的

ventilation 通风,流通空气

grill 烤架,铁格子

entomology 昆虫学

porous 多孔渗水的

stainless 不锈的

seam 接缝,线缝

bolt 门闩,螺钉

nut 螺母,螺帽

washer 垫圈

gasket 垫圈,衬垫

motor 发动机,电动机

pulley 滑车,滑轮

drum 鼓形圆桶

clean in place (CIP) 现场清洗系统

clean out of place (COP) 拆卸清洗

sluice 水闸,泄水;开闸放水,流出,冲洗,奔泻

abrasive 研磨剂;研磨的

adulteration 掺杂,掺假货

Exercises

Ⅰ. **Answer the following questions according to the article.**

1. What is GMP? Which areas are covered for it?

2. How to keep the personal cleanliness?

3. What are requires of personal frocks?

4. How to control personal hair?

5. When do personal wash their hands?

6. How to clean equipments?

7. What is SSOP? There are eight topics in SSOP, what are they?

8. Which conducts do not permit for personal in workshop?

Ⅱ. **Choose a term from what we have learnt to fill in each of the following blanks. Change the word form where necessary.**

1. GMPs is the abbreviation of _____.

2. GMPs including buildings, _____, equipment, production and process controls, _____,

and distribution.

3. SSOPs is the abbreviation of _____.

4. Food contact surfaces should be _____.

5. All food operations must follow sanitary principles, such as receiving, _____, packaging, _____, processing, _____.

6. The most important aspect in sanitation is the commitment to producing products in a clean plant environment.

7. The placement of _____ has a direct impact on ease of cleaning and accessibility.

8. Grass or weeds around the facility should be mowed or kept short to eliminate _____ _____.

Unit 4 HACCP

Hazard Analysis and Critical Control Points (HACCP) is a logical, scientific approach to controlling hazards in food production. HACCP is a preventive system for assuring the safe production of food products. The application of HACCP is based on technical and scientific principles that assure food safety. An ideal application would include all processes from the farm to the table. The principle of HACCP can be applied to production, meat slaughter and processing, shipping and distribution, food service and in home preparation.

The HACCP system, which is science based and systematic, identifies specific hazards and measures for their control to ensure the safety of food. HACCP is a tool to assess hazards and establish control systems that focus on prevention rather than relying mainly on end-product testing and inspection. Any HACCP system is capable of accommodating change, such as advances in equipment design, processing procedures or technological developments.

4.1 Principles of The HACCP System

The HACCP system consists of the following seven principles:

Principle 1: Conduct a Hazard Analysis

Identify the potential hazard(s) associated with food production at all stages, from primary production, processing, manufacture and distribution until the point of consumption. Assess the likelihood of occurrence of the hazard(s) and identify the measures for their control.

Principle 2: Determine the Critical Control Points (CCPs)

Determine the points, procedures or operational steps that can be controlled to eliminate the hazard(s) or minimize its (their) likelihood of occurrence.

A "step" means any stage in food production and/or manufacture including the receipt and/or production of raw materials, harvesting, transport, formulation, processing, storage, etc.

Principle 3: Establish Critical Limit(s)

Establish critical limit(s) which must be met to ensure the CCP is under control.

Principle 4: Establish Monitoring Procedures

Establish a system to monitor control of the CCP by scheduled testing or observations.

Principle 5: Establish Corrective Actions

Establish the corrective action to be taken when monitoring indicates that a particular CCP is not under control.

Principle 6: Establish Verification Procedures

Establish procedures for verification to confirm that the HACCP system is working effectively.

Principle 7: Establish Recordkeeping Procedures

Establish documentation concerning all procedures and records appropriate to these principles and their application.

4.2　Executing a HACCP Plan

The five preliminary steps are:

Step 1: Assemble the HACCP Team

The first step is to assemble the HACCP resources and team. For the development of a HACCP plan a meat processing company should bring together as much knowledge as possible. Companies should assemble written materials and company documents that relate to food safety as well as assemble a team of individuals that represent different segments within the industry. One HACCP coordinator with HACCP skills should be appointed. HACCP skills are not necessary for other members to be on team. The team should be multi-disciplinary and represent all areas of plant such as engineering, production, sanitation, and quality assurance. Some companies may consider including outside experts from Universities or trade associations. Larger companies may develop teams of seven or eight people while small companies may have teams as small as two or three people.

The HACCP coordinator should have overall responsibility for the HACCP program and should play the role of team leader. The HACCP coordinator must have management skills, must be trained in the HACCP principles and needs to have the company resources to implement HACCP.

Step 2: Describe the Product and its Method of Distribution

The second step is to describe completely each food product the plant makes. This can include a brief description of how the process occurs and/or how the product is produced or prepared. This will help to focus on potential hazards that may occur in the product. To describe the product the company should answer the questions in the form below.

Product Category Description

The following areas need to be defined when developing the product category description:

1. Common Name/Description:

2. Process Description:

3. How is it to be used?

4. Type of Package?

5. Length of Shelf Life; at what temperature?

6. Where will it be sold?

7. Labeling instructions;

8. Is special distribution control needed?

Step 3: Develop a Complete List of Ingredients and Raw Materials

The third step is to develop a written list of ingredients and raw materials for each process/product. The ingredients and raw materials will help to focus on potential hazards in the meat product produced. Some processors have found it helpful to divide the ingredients as indicated in the form below.

Product and Ingredients

Product:

Meat Ingredients:

Non-Meat Ingredients:

Restricted Ingredients:

Packaging Materials:

Casing:

Step 4: Develop a Process Flow Diagram

The next step is to construct a process flow diagram that identifies all the steps used to prepare the product, from receiving through final shipment, that are directly under the control of the processing plant. After the flow diagram is constructed it should be verified by walking through the plant to make sure that the steps listed on the diagram describe what really occurs in producing the product.

Step 5: Meet the Regulatory Requirements for Sanitation

Good sanitation is the most basic way to ensure that a safe product is produced. Maintaining good sanitation serves as an excellent foundation for building a HACCP plan. It also demonstrates that plant management has the commitment and resources to successfully implement the HACCP plan. The regulatory requirement for Sanitation Standard Operating Procedures is a pre-HACCP.

Application of the Principles of HACCP

Step 6: Conduct a Hazard Analysis

The application of this principle involves listing the steps in the process and identifying where significant hazards are likely tooccur. The HACCP team will focus on hazards that can be prevented, eliminated or controlled by the HACCP plan. A justification for including or excluding the hazard is reported and possible control measures are identified.

Step7: Identify the Critical Control Points

A critical control point (CCP) is a point, step or procedure at which control can be applied and a food safety hazard can be prevented, eliminated or reduced to acceptable levels. The HACCP team will use a CCP decision tree to help identify the critical control points in the process. A critical control point may control more than one food safety hazard or in some cases more than one CCP is needed to control a single hazard. The number of CCP's needed depends on the processing steps and the control needed to assure food safety.

Step 8: Establish Critical Limits

A critical limit (CL) is the maximum and/or minimum value to which a biological, chemical, or physical parameter must be controlled at a CCP to prevent, eliminate, or reduce to an acceptable level the occurrence of a food safety hazard. The critical limit is usually a measure such as time, temperature, water activity (Aw), pH, weight, or some other measure that is based on scientific literature and/or regulatory standards.

Step 9: Monitor CCP

The HACCP team will describe monitoring procedures for the measurement of the critical limit at each critical control point. Monitoring procedures should describe how the measurement will be taken, when the measurement is taken, who is responsible for the measurement and how frequently the measurement is taken during production.

Step10: Establish Corrective Action

Corrective actions are the procedures that are followed when a deviation in a critical limit occurs. The HACCP team will identify the steps that will be taken to prevent potentially hazardous food from entering the food chain and the steps that are needed to correct the process. This usually includes identification of the problems and the steps taken to assure that the problem will not occur again.

Step11: Verification

Those activities, other than monitoring, that determine the validity of the HACCP plan and that the system is operating according to the plan. The HACCP team may identify activities such as auditing of CCP's, record review, prior shipment review, instrument calibration and product testing as part of the verification activities.

Step12: Record keeping

A key component of the HACCP plan is recording information that can be used to prove that the food was produced safely. The records also need to include information about the HACCP plan. Record should include information on the HACCP Team, product description, flow diagrams, the hazard analysis, the CCP's identified, Critical Limits, Monitoring System, Corrective Actions, Recordkeeping Procedures, and Verification Procedures.

New Words

Hazard Analysis and Critical Control Points (HACCP) 危害分析与关键控制点
slaughter   屠宰，残杀，屠杀
hazard analysis   危害分析
Critical Control Points (CCPs)   关键控制点
critical limit(s)   关键限值
monitoring procedures   监控程序
corrective actions   纠偏措施（行动）
verification procedures   验证程序
recordkeeping procedures   记录保持程序
preliminary   预备的，初步的

coordinator　协调者，同等的人或物
multidisciplinary　包括各种学科的，有关各种学问的
category　种类，别，【逻】范畴
process flow diagram　过程流程图
verification　确认，查证，验证

Exercises

Ⅰ. Answer the following questions according to the article.

1. What is the definition of HACCP?
2. What are the principles of HACCP system?
3. How todetermine CCPs?
4. What are the preliminary steps for executing a HACCP plan?
5. How todescribe a food product?

Ⅱ. Choose a term from what we have learnt to fill in each of the following blanks. Change the word form where necessary.

1. The roles of HACCP are _____.
2. The _____ will help to focus on potential hazards in the meat product produced.
3. A critical control point is a _____, _____ or _____ at which control can be applied and a food safety hazard can be prevented, eliminated or reduced to acceptable levels.
4. The critical limit is usually a measure such as _____, _____, _____, _____, _____, or some other measure that is based on scientific literature and/or regulatory standards.
5. Corrective actions are the procedures that are followed when a _____ in a critical limit occurs.

Unit 5　Food Risk Analysis

5.1　Risk Analysis

There are many hazards associated with food that can and do result in injury and harm to human health. Millions of people worldwide suffer from some sort of "food poisoning" each year. Uncontrolled application of agricultural chemicals, environmental contamination, use of unauthorized additives, microbiological hazards and other abuses of food along the food chain can all contribute to the potential of introducing or failing to reduce hazards related to food. With increased awareness of the effects of food hazards on human health, the increasing importance and rapid growth of world food trade and the demand by consumers for a safe food supply, analysis of the risks associated with food has become more important than ever before.

Consumers have expressed concern about the safety of food additives, agricultural and veterinary chemical residues, biological, chemical and physical contaminants, radionuclide contamination and uncontrolled and unacceptable food handling practices and processing which can result in the introduction of hazards to food at all stages along the food chain, from primary production to the consumer. Consumers in the developed world have voiced these concerns most often; however, continuous improvements in global communication have heightened the interest of consumers throughout the world on these matters.

Process of Risk Analysis

The risk to the world's population from hazards in and on food depends largely on the degree of control exercised by producers, processors and official food control authorities to prevent or minimize the risks to acceptable safe levels. Food safety risk analysis is an emerging discipline, and the methods used for assessing and managing risks associated with food hazards are still being developed.

It is important to recognize the difference between "hazard" and "risk". As stated above, a hazard is a biological, chemical or physical agent in, or condition of, food with the potential to cause harm. In contrast, risk is the estimated probability and severity of adverse health effects in exposed populations consequential to hazards in food. Understanding the association between a reduction in hazards that may be associated with a food and the reduction in the risk of adverse health effects to consumers is of particular importance in development of appropriate food safety controls. Unfortunately, there is no such thing as "zero risk" for food (or for anything else).

The risk analysis process comprises three separate elements: risk assessment, risk management and risk communication. It is widely recognized as the fundamental methodology underlying the development of food safety standards. Decisions are needed to determine what the hazards are and to identify their immediate, interim and long-term effects on human health (risk assessment); to establish the appropriate measures of control to prevent, reduce or minimize these risks (risk management); and to determine the best way to communicate this information to the affected population (risk communication).

5.2 Risk Assessment

Risk assessment is a quantitative evaluation of information on potential health hazards from exposure to various agents. It involves four interrelated steps:

- Identification of the hazard and comprehension of the danger it represents, the impact in terms of human health and the circumstances under which the danger is present (hazard identification)
- Qualitative and/or quantitative evaluation of the adverse effects of the hazard on human health (hazard characterization).
- Qualitative and/or quantitative evaluation of the likely degree of consumption or intake of the hazardous agent (exposure assessment).
- Integration of the first three steps into an estimate of the likely adverse effect in the target population (risk characterization).

The entire risk assessment process requires the use of sound and scientifically derived information and the application of established scientific procedures carried out in a transparent manner. Unfortunately, sound scientific data are not always available for the qualitative and quantitative evaluations necessary for an absolutely sure final decision; consequently a degree of uncertainty must be factored into the decision.

The importance of risk assessment lies not only in its capacity for estimating human risk, but also in its function as a framework for organizing data as well as for allocating responsibility for analysis. The risk assessment process can include a variety of models for reaching conclusions; for example, the concept of acceptable daily intake (ADI) may be considered a component of risk assessment.

Biological hazards of concern to public health include pathogenic strains of bacteria, viruses, helminthes, protozoa, algae and certain toxic products they may produce. Of these hazards, pathogenic bacteria in foods currently present the most significant problems internationally. Assessment of the risks associated with bacterial pathogens presents unique complications. Any method for assessing the risk of hazards from food-borne bacteria will be complicated by factors related to methods used to grow, process and store food for consumption. These factors can vary greatly depending on cultural and geographical differences. Such factors characterize the scenario for a given food and are an essential element for a risk assessment for bacterial hazards.

In many cases sufficient data will not be available to support a quantitative assessment of risks associated with bacterial pathogens. For a number of reasons, including the many uncertainties associated with how and when an organism may express its pathogenic potential, it has not yet been determined whether a quantitative risk assessment approach is possible and appropriate for characterization of risk associated with food-borne bacterial pathogens. Thus, by default, a qualitative approach to characterizing risk may be the only current alternative. To bring about regulatory changes the scientific community must advance beyond qualitative microbial risk assessment and generate the data needed to make quantitative assessments. FAO/WHO consultations had difficulty with quantitative microbiological risk assessment, and one recommendation is to establish an FAO/WHO Expert Committee on Microbiological Risk Assessment.

Chemical risk assessment is a fairly well established process and in general permits the assessment of risks from long-term chronic exposure to a chemical. It includes the assessment of food additives, residues of pesticides and other agricultural chemicals, residues from veterinary drugs, chemical contaminants from any source and natural toxins such as mycotoxins and ciguatoxin.

Risk assessment requires evaluation of relevant information and selection of the models to be used in drawing inferences from that information. Further, it requires recognition of uncertainties and, when appropriate, acknowledgement that alternative interpretations of the available data may be scientifically plausible. Data uncertainties arise both from limitations on the amount of data available and from evaluation and interpretation of actual data obtained from epidemiological and toxicological studies. Model uncertainties arise whenever attempts are made to use data concerning phenomena that are likely to occur under other sets of conditions for which data are not available.

5.3 Risk Management

Risk management is defined within the Codex Alimentarius as the process of weighing policy alternatives in the light of the results of risk assessment and, if required, selecting and implementing appropriate control options including regulatory measures. The goal of the risk management process is to establish the significance of the estimated risk, to compare the costs of reducing this risk to the benefits gained, to compare the estimated risks to the societal benefits derived from incurring the risk and to carry out the political and institutional process of reducing the risk.

Elements of risk management include the following:

A. Risk evaluation
- Identification of a food safety problem.
- Establishment of a risk profile.
- Ranking of the hazard for risk assessment and risk management priority.
- Establishment of risk assessment policy for conduct of risk assessment.
- Commissioning of risk assessment.
- Consideration of risk assessment result.

B. Risk management option assessment
- Identification of available management options.
- Selection of preferred management option, including consideration of an appropriate safety standard.
- Final management decision.

C. Implementation of management decision

D. Monitoring and review
- Assessment of effectiveness of measures taken.
- Review risk management and/or assessment as necessary.

The outcome of the risk management process, as undertaken by committees within the Codex Alimentarius system, is the development of standards, guidelines and other recommendations for food safety. In the national situation it is likely that different risk management decisions could be made according to different criteria and different ranges of risk management options. Risk managers, in developing approaches to managing risk, use the risk characterization that results from the risk assessment process. Risk management decisions can be based on establishing safe handling procedures and practices, food processing quality and safety assurance controls and food quality and safety standards to control hazards in food. These standards must take into consideration the proper use of food additives which have been determined to be safe and their permitted levels and scientifically determined acceptable safe limits for contaminants and agricultural chemical residues in food, using the risk assessment process.

The outcome of the risk assessment process should be combined with the evaluation of available risk management options in order that a decision on management of the risk can be reached. Implementation of the management decision should be followed by monitoring of both the effectiveness of the control measure and its impact on the risk to the exposed consumer population, to ensure that the food safety objective is being met.

While research and scientific studies continue to provide the answers needed for making informed decisions in risk analysis related to hazards in food, the uncertainty and unresolved questions continue to cause concern to decision-makers. Only continued research and scientific study can provide the necessary answers. Until these answers are available, much of what is known about hazards and assessing and controlling risks is based on only partial information, with uncertainties factored into the analysis.

General principles of food safety risk management

Principle 1: Risk management should follow a structured approach

The elements of a structured approach to risk management are Risk Evaluation, Risk Management Option Assessment, Implementation of Management Decision, and Monitoring and Review. In certain circumstances, not all of these elements will be included in risk management activities (e.g. standard setting by Codex, with implementation of control measures by national governments).

Principle 2: Protection of human health should be the primary consideration in risk management decisions

Decisions on acceptable levels of risk should be determined primarily by human health considerations, and arbitrary or unjustified differences in the risk levels should be avoided. Consideration of other factors (e.g. economic costs, benefits, technical feasibility, and societal preferences) may be appropriate in some risk management contexts, particularly in the determination of measures to be taken. These considerations should not be arbitrary and should be made explicit.

Principle 3: Risk management decisions and practices should be transparent

Risk management should include the identification and systematic documentation of all elements of the risk management process including decision-making, so that the rationale is transparent to all interested parties.

Principle 4: Determination of risk assessment policy should be included as a specific component of risk management

Risk assessment policy sets the guidelines for value judgements and policy choices which may need to be applied at specific decision points in the risk assessment process, and preferably should be determined in advance of risk assessment, in collaboration with risk assessors.

Principle 5: Risk management should ensure the scientific integrity of the risk assessment process by maintaining the functional separation of risk management and risk assessment

Functional separation of risk management and risk assessment serves to ensure the scientific integrity of the risk assessment process and reduce any conflict of interest between risk assessment and risk management. However, it is recognized that risk analysis is an iterative process, and interactions between risk managers and risk assessors are essential for practical application.

Principle 6: Risk management decisions should take into account the uncertainty in the output of the risk assessment

The risk estimate should, wherever possible, include a numerical expression of uncertainty, and this must be conveyed to risk managers in a readily understandable form so that the full implications of the range of uncertainty can be included in decision-making. For example, if the risk estimate is highly uncertain the risk management decision might be more conservative.

Principle 7: Risk management should include clear, interactive communication with consumers and other interested parties in all aspects of the process

On-going reciprocal communication among all interested parties is an integral part of the risk management process. Risk communication is more than the dissemination of information, and a major function is the process by which information and opinion essential to effective risk management is incorporated into the decision.

Principle 8: Risk management should be a continuing process that takes into account all newly generated data in the evaluation and review of risk management decisions

Subsequent to the application of a risk management decision, periodic evaluation of the decision should be made to determine its effectiveness in meeting food safety objectives. Monitoring and other activities will likely be necessary to carry out the review effectively.

5.4 Risk Communication

Risk communication is the third and final element of the risk analysis process. The Codex Alimentarius definition of risk communication is narrow: "an interactive process of exchange of information and opinion on risk among risk assessors, risk managers and other interested parties". A definition with broader scope is that of the United States National Academy of Sciences: "an interactive process of exchange of information and opinion among individuals, groups and institutions which involves multiple messages about the nature of risk and other messages, not strictly about risk, that express concerns, opinions or reactions to risk messages or to legal and institutional arrangements for risk management".

Communicating the results of risk assessment and risk management serves many purposes. The quality and safety of food depends on responsible action by all involved at all stages in the food chain, including consumers. Consumers require access to adequate information about potential hazards and appropriate precautions to be taken in the final preparation and serving of food. In addition, consumers need to be aware of and to understand food safety control measures implemented by their government in the interest of consumers' health.

Communication provides the public with the results of expert scientific review of food hazard identification and assessment of the risks to the general population or to specific target groups such as infants or the elderly. Certain people, such as those who are immuno-deficient, allergic or nutritionally deficient, require particular information. Communication provides the private and public sectors with the information necessary for preventing, reducing or minimizing food risks to acceptably safe levels through systems of food quality and safety management by either mandatory or voluntary means. It also provides sufficient information to permit the populations with the greatest level of risk from any particular hazard to exercise their own options for achieving even greater levels of protection.

New Words

risk analysis 风险分析
food poisoning 食物中毒
food additive 食品添加剂
agricultural and veterinary chemical residues 农药、兽药残留
radionuclide 放射性核
zero risk 零风险
risk assessment 风险评估
risk management 风险管理
risk communication 风险交流
methodology 方法学，方法论
interim 中间的，临时的，间歇的；中间时期，过渡时期，暂定
hazard identification 危害识别
hazard characterization 危害描述
exposure assessment 暴露评估
risk characterization 风险描述
qualitative 性质上的，定性的
quantitative 数量的，定量的
helminth 寄生虫，蠕虫
protozoa 原生动物
algae 藻类，海藻
risk evaluation 风险评估
risk management option assessment 风险管理选择评估
implementation of management decision 执行风险决定
monitoring and review 监测和评价

Exercises

I. Answer the following questions according to the article.

1. Which factors can cause food poisoning? Please write out five factors.
2. What are the difference between "hazard" and "risk"?
3. Which elements are composed of the risk analysis process?
4. What is the definition of food assessment?
5. What are the steps of risk assessment?
6. What are the difficulties in therisk assessment process?
7. What are the elements of risk management?
8. What are the principles of food safety risk management?

Ⅱ. **Choose a term from what we have learnt to fill in each of the following blanks. Change the word form where necessary.**

1. In order to prevent or minimize the risks to acceptable safe levels, food safety depends largely on _____.

2. Risk is the estimated _____ and _____ of adverse health effects in exposed populations consequential to hazards in food.

3. The risk analysis process comprises three separate elements: _____, _____ and _____.

4. Risk assessment is a _____ evaluation of information on potential health hazards from exposure to various agents.

5. The entire risk assessment process requires the use of _____ and _____ derived information and the application of established scientific procedures carried out in a _____ manner.

6. Of food hazards, _____ in foods currently present the most significant problems internationally.

7. _____ risk assessment is a fairly well established process and in general permits the assessment of risks from _____ exposure to a chemical.

8. The third and final element of the risk analysis process is _____.

Unit 6　Food Quality

6.1　Definition of Food Quality

In the last decades quality has become of utmost importance to society. Consumers have become more conscious of quality, and organizations are now judged more on their overall quality performance instead of their financial performance alone.

Quality is defined by the International Organization for Standardization (ISO) as "the totality of features and characteristics of a product that bear on its ability to satisfy stated or implied needs." In other words, good quality exists when the product complies with the requirements specified by the client. This means quality is a term defined by the consumer, buyer, grader, or any other client based on a number of subjective and objective measurements of the food product. These may include measures of purity, flavor, color, maturity, safety, wholesomeness, nutrition, or any other attribute or characteristic of the product.

In the marketing and consumer economics literature, two main approaches are taken to define food quality. The holistic approach includes within the concept of food quality "all the desirable characteristics a product is perceived to have". By contrast, the excellence approach views food quality as referring only to characteristics that pertain to a higher, more restrictive or "superior" specification of the product. The holistic approach has been adopted in this report to cover the whole range of quality assurance schemes currently existing.

The holistic approach leaves wide scope for interpretation: quality can mean conforming to standards (including standards pertaining to the environment, local specialties, organic production, ethics, and even taste and smell) and it can refer to subjectively perceived quality attributes.

Quality is also a factor that involves the entire production process, from raw materials, processing and packaging up to consumption of the product. As the idea of quality is continuously evolving, any attempts to classify it are immediately overtaken by events, as new meanings are added to existing ideas without replacing them. It is helpful to simplify the different quality requirements into two categories——"musts" and "wants":

(1) *musts* have to be present in order for the product or service to be assessed as acceptable;

(2) *wants* depend on the wishes or expectations that influence choices.

In the past quality was mainly a question of "musts", whereas nowadays it also includes a large proportion of "wants".

6.2 Quality Attributes

Food product attributes can be grouped into search attributes, experience attributes and credence attributes:

"**Search attributes**" are characteristics that can be identified and recognized from the outside before choosing the product (look, price, variety, etc.).

"**Experience attributes**" are characteristics which are not directly perceivable when the product is chosen, but become so when it is consumed (tasty, solid, easy/quick to prepare, etc.) and prompt users to decide whether or not they will consume that product again.

"**Credence attributes**" are characteristics which are not perceivable when the product is purchased or consumed, and which users cannot personally and directly assess. For this group of attributes the trust required of users becomes fundamental, as does the role of information to bridge this sort of "knowledge gap" on the part of consumers. The credence attributes category includes all the characteristics related to places and methods of production, use of certain substances and, in a broad sense, the level of safety associated with the product.

6.3 Quality Factors in Foods

Quality of a food product involves maintenance or improvement of the key attributes of the product—including color, flavor, texture, safety, healthfulness, shelf life, and convenience. To maintain quality, it is important to control microbiological spoilage, enzymatic degradation, and chemical degradation. These components of quality depend upon the composition of the food, processing methods, packaging, and storage.

6.3.1 Appearance Factors

Color

Color is a very important appearance factor. Consumers expect meat to be red, apple juice to be light brown and clear, orange juice to be orange, egg yolks to be bright yellow-orange, and so on. Food color measurements provide an objective index of food quality. Color is an indication of ripeness or spoilage. The end point of cooking processes is judged by color. Changes in expected colors can also indicate problems with the processing or packaging.

Browns and blackish colors can be either enzymatic or non-enzymatic reactions. The major non-enzymatic reaction of greatest interest to scientists is the Maillard reaction, which is the dominant browning reaction. Other less explained reactions include blackening in potatoes or the browning in orange juice. The enzymatic browning found widespread in fruits and selected vegetables is due to the enzymatic catalyzed oxidation of the phenolic compounds.

Naturally occurring pigments play a role in food color. Water-soluble pigments may be categorized as anthocyanins and anthoxthanins. Lesser known water-soluble pigments include the leucoanthocyanins. Fat-soluble plant pigments are primarily categorized into the chlorophyll and carotenoid pigments. These green and orange-yellow pigments considerably impact the color. Myoglobins contribute to the color of meat.

Size and Shape

Depending on the product, consumers expect foods to have certain sizes and shapes. For example, consumers have some idea of what an ideal French fry should look like, or an apple, or a cookie, or a pickle. Size and shape are easily measured. Fruits and vegetables are graded based on their size and shape, and this is done by the openings they will pass through during grading. Now computerized electronic equipment can determine the size and shape of foods.

6.3.2 Textural Factors

Consumers expect gum to be chewy, crackers to be crisp, steak to be tender, cookies to be soft, and breakfast cereal to be crunchy. The texture of food refers to the qualities felt with the fingers, the tongue, or the teeth. Textures in food vary widely, but any departure from what the consumer expects is a quality defect.

Texture is a mechanical behavior of foods measured by sensory (physiological/psychological) or physical (rheology) means. Rheology is the study of the science of deformation of matter. The four main reasons for studying rheology include:

1. Insight into structure
2. Information used in raw material and process control in industry
3. Applications to machine design
4. Relation to consumer acceptance

Regardless of the reason for studying texture, classification and understanding are difficult because of the enormous range of materials. Moreover, food materials behave differently under different conditions.

Changes in texture are often due to water status. Fresh fruits and vegetables become soggy as cells break down and lose water. On the other hand if dried fruits take on water, their texture changes. Bread and cake lose water as they become stale. If crackers, cookies, and pretzels take up water, they become soft and undesirable.

Various methods are used to control the texture of processed foods. Lipids (fats) are softeners and lubricants used in cakes. Starch and gums are used as thickeners. Protein can also be a thickener, or if coagulated as in baked bread, it can form a rigid structure. Depending on its concentration in a product, sugar can add body as in soft drinks or in other products add chewiness, or in greater concentrations it can thicken and add chewiness or brittleness.

6.3.3 Flavor Factors

Food flavor includes taste sensations perceived by the tongue—sweet, salty, sour, and bitter—and smells perceived by the nose. Often the terms flavor and smell (aroma) are used interchangeably.

Food flavor and aroma are difficult to measure and difficult to get people to agree on. A part of food science called sensory science is dedicated to finding ways to help humans accurately describe the flavors and other sensory properties of their food.

Flavor, like color and texture, is a quality factor. It influences the decision to purchase and to consume a food product. Food flavor is a combination of taste and smell, and it is very subjective and difficult to measure. People differ in their ability to detect tastes and odors. People also differ in their preferences for these.

Besides the tastes of sweet, salty, sour, and bitter, an endless number of compounds give food characteristic aromas, such as fruity, astringency, sulfur, hot.

Sweetness may result from sugars like arabinose, fructose, galactose, glucose, riboses, xylose, and other sweeteners. Organic acids may be perceived on the bottom of the tongue. Some of these common acids are citric, isocitric, malic, oxalic, tartaric, and succinic acids. The fruity flavors are often esters, alcohols, ethers, or ketones. Many of these are volatile and are associated with acids.

Phenolic compounds are closely related to the sensory and nutritional qualities of plants. They are found in many fruits, including apples, apricots, peaches, pears, bananas, and grapes; and vegetables such as avocado, eggplant, and potatoes, and contribute to color, astringency, bitterness, and aroma. Most phenoliccompounds are found around the vascular tissues in plants, but they have the potential to react with other components in the plant as damage to the structure occurs during handling and processing. Loss of nutrients and changes in color and flavor occur in foods due to the phenolic compounds' reaction with polyphenol oxidase, or PPO, an enzyme that catalyzes oxidation.

The sense of taste is a powerful predictor of food selection. The four main tastes the body experiences are sweet, sour, salty, and bitter. Humans like sweet-tasting foods. Possibly this preference for sweet is a holdover from ancient ancestors, who found that sweetness indicated that the food provided energy.

Judgment of flavor is often influenced by color and texture. Flavors such as cherry, raspberry, and strawberry are associated with the color red. Beef flavor is brown. Actually, the flavor essences are colorless. As for texture, people expect potato chips to be crunchy and gelatin to be soft and cool. Depending on the food, flavor can also be influenced by: Bacteria, Yeasts, Molds, Enzymes, Heat/cold, Moisture/dryness, Light, Time, Additives.

Finally, depending on the product, the influence these factors have on a food flavor can be positive or negative and sometimes differs depending on the person.

6.3.4 Additional Quality Factors

Additional quality factors include shelf life, safety, healthfulness, and convenience. The extension of storage life of products generally involves heat treatments, irradiation, refrigeration or freezing, or reduction of water activity by either addition of water-binding agents, like sugars, or drying. In many cases compromises are made to achieve desired shelf life or convenience. Such processes, though improving shelf life, almost always have some effect on the components of the food. The factors that influence changes of various ingredients in foods include the following: proteins, lipids, carbohydrates, vitamins, chemicals, and microbiological characteristics.

Proteins

Heat denaturation changes solubility and texture of foods; light oxidation of protein causes off flavors. Enzymatic degradation of protein can cause changes in body and texture and also bitter flavors.

Freezing can alter protein conformation and solubility in some cases.

Lipids

Enzymatic hydrolysis of lipids can cause off flavors, such as soapy or goaty, depending on type of oil. This also makes frying oilsunsuitable for use. It can change functionality and crystallization properties. Oxidation of unsaturated fatty acids causes off flavors.

Carbohydrates

High-heat treatments cause interactions between reducing sugars and amino groups to give Maillard browning and changes in flavor. Hydrolysis of starch and gums can change texture of food systems. Some starches can be degraded by enzymes or under acidic conditions.

Vitamins

Depending upon the vitamin, losses can occur when the food is heated, exposed to light, or to oxygen.

Chemicals and Microbiological Characteristics

Ensuring the safety of food involves careful control of the process from the farm gate to the consumer. Safety includes control of both chemical and microbiological characteristics of the product. Most processing places emphasis on microbial control, and often has as its objective the elimination of organisms or prevention of their growth.

Processes that are aimed at prevention of growth include:

—Irradiation

—Refrigeration

—Freezing

—Drying

—Control of water activity (addition of salt, sugars, polyols, and so forth)

Processes that are aimed at minimizing organisms include:

—Pasteurization

—Sterilization (canning)

—Cleaning and sanitizing

—Membrane processing

A further method of processing that is aimed at the control of undesirable microflora is the deliberate addition of microorganisms and the use of fermentation.

Safety from a chemical viewpoint generally relates to keeping undesirable chemicals, such as pesticides, insecticides, and antibiotics, out of the food supply. Making sure that food products arefree from extraneous matter (metal, glass, wood, etc.) is another facet of food safety.

Today's consumers want food products that are convenient to use and still have all the qualities of a fresh product.

New Words

maturity　成熟，完备，(票据)到期，成熟
wholesome　卫生的，有益的，健康的，有益健康的
holistic　整体的，全盘的
pertain　适合，属于
search attribute　搜索特性
experience attribute　体验特性
credence attribute　信誉特性
spoilage　损坏，腐败
enzymatic　【生化】酶的
degradation　降级，降格，退化
yolk　蛋黄，【生物】卵黄
ripen　使成熟；成熟
Maillard reaction　美拉德反应
phenolic　【化】酚的，石碳酸的
pigment　【生】色素，颜料
anthocyanin　【植】花青素，【化】花色醣苷
leucoanthocyanin　无色花色苷
chlorophyll　【生化】叶绿素
carotenoid　【生化】类胡萝卜素
myoglobin　肌球素，肌血球素
pickle　(腌肉、菜等用的)盐汁，醋汁，腌渍品，泡菜
gum　口香糖，香口胶，泡泡糖
chewy　(食物)耐嚼的，不易嚼碎的
crisp　脆的，易碎的
rheology　【物】流变学，流变能力
lubricant　滑润剂
starch　淀粉

aroma　芳香，香气，香味
fruity　果味的，（酒）有葡萄味的
astringency　收敛性，严酷
arabinose　【化】树胶醛醣
fructose　果糖
galactose　【生化】半乳糖
glucose　葡萄糖
ribose　【生化】核糖
xylose　【化】木糖，戊醛糖
citric　柠檬的
isocitric　异柠檬的
malic　苹果的，由苹果取得的
oxalic　酢浆草的，采自酢浆草的
tartaric　【化】酒石的，似酒石的，含有酒石的
ester　【化】酯
alcohol　酒精，酒
ether　醚
ketone　【化】酮
avocado　【植】鳄梨，鳄梨树
vascular　【解】【动】脉管的，有脉管的，血管的
polyphenol oxidase　多酚氧化酶
gelatin　凝胶，白明胶
denaturation　改变本性，使变性
soapy　碱味
goaty　膻味
crystallization　结晶化
pasteurization　加热杀菌法，巴斯德杀菌法
sterilization　杀菌

Exercises

I．Answer the following questions according to the article.

1. What are the definition of quality and food quality?
2. Please give several examples of search attributes, experience attributes and credence attributes about foods.
3. How many quality factors in foods? Please write them out.
4. How many favors in foods?
5. What are the additional quality factors?
6. What factors do influence changes of various ingredients in foods?

7. There are many methods of minimizing organisms, please tell us three ones.

8. What is Maillard reaction? Please write out the process of Maillard reaction.

Ⅱ. Choose a term from what we have learnt to fill in each of the following blanks. Change the word form where necessary.

1. Good quality exists when the product _____ the requirements specified by the client.

2. There are two main approaches to define food quality, they are _____ and _____.

3. It is helpful to simplify the different quality requirements into two categories—"_____" and "_____". In the past quality was mainly a question of "_____", whereas nowadays it also includes a large proportion of "_____".

4. Food product attributes can be grouped into _____, _____ and _____.

5. Appearance factors in foods include _____, _____ and _____.

6. The texture of food refers to the qualities felt with _____, _____ or _____.

7. Food flavor includes _____ perceived by the tongue and _____ perceived by the nose.

8. Maillard reaction is a reaction between _____ and _____.

Unit 7　Food Quality Management

Food Quality Management focuses on consumer-driven Quality Management in food production systems using a product-based approach. It integrates organizational and technological aspects of food product quality into one techno-managerial concept and it presents an integrated view of how Quality Management is to be situated in a chain-oriented approach.

7.1　Functions of Quality Management

Management is creative problem solving. This creative problem solving is accomplished through five functions of management: planning, organizing, staffing, directing and controlling. The intended result is the use of an organization's resources in a way that accomplishes its mission and objectives.

Planning is the ongoing process of developing the business´ mission and objectives and determining how they will be accomplished. Planning includes both the view of the organization, e.g., its mission, and the narrowest, e.g., a tactic for accomplishing a specific goal.

Organizing is establishing the internal organizational structure of the organization. The focus is on division, coordination, and control of tasks and the flow of information within the organization. It is in this function that managers distribute authority to job holders.

Staffing is filling and keeping filled with qualified people all positions in the business. Recruiting, hiring, training, evaluating and compensating are the specific activities included in the function. In the family business, staffing includes all paid and unpaid positions held by family members including the owner/operators.

Directing is influencing people's behavior through motivation, communication, group dynamics, leadership and discipline. The purpose of directing is to channel the behavior of all personnel to accomplish the organization's mission and objectives while simultaneously helping them accomplish their own career objectives.

Controlling is a process of establishing performance standards based on the firm's objectives, measuring and reporting actual performance, comparing the two, and taking corrective or preventive action as necessary.

Each of these functions involves creative problem solving. Creative problem solving is broader than problem finding, choice making or decision making.

7.2　Principles of Quality Management

Organizations that focus on quality rely on the same basic management principles for success, no matter what kind of product or service they provide. These principles are:

- **Strengthen systems and processes.** By viewing an organization as a collection of interdependent systems and processes, managers can understand how problems occur and can strengthen the organization as a whole.
- **Encourage staff participation and teamwork.** Every employee can help assure good quality if managers empower staff members to solve problems and recommend improvements.
- **Base decisions on reliable information.** By collecting and analyzing accurate, timely, and objective data, managers can diagnose and solve organizational problems and measure progress.
- **Improve communication and coordination.** Different units, facilities, and management levels can work together to improve quality if they share information freely and coordinate their activities.
- **Demonstrate leadership commitment.** When top leaders are committed to good quality, employees accept it as a guiding principle for their own work.

7.3　Quality Design

Managers and planners can help assure good care and prevent problems from arising by designing quality into every aspect of a program, including its mission and objectives, allocation of resources, and development of standards and guidelines. Quality design is one point of the quality assurance triangle and a prerequisite for the other two points, quality control and quality improvement.

Mission and Objectives

Quality design begins with defining the organization's mission, including its purpose, values, objectives, and clients, with an eye to quality. This is the first step both for creating a new program and for redesigning an existing program. To develop realistic objectives concerning quality, managers must clearly assess the level of quality that can be achieved with available resources; the institution's strengths and weaknesses, including current program performance and quality; the client population, including how clients themselves perceive quality; and the political and social climate. Objectives foster quality best when they focus on meeting clients' needs.

In quality design the participation of front-line service providers, field supervisors, and clients is crucial. Top decision-makers rarely have direct experience with day-to-day service delivery. Without insights from providers and clients, intended improvements may not be meaningful to the staff who must implement them and may not attract clients or meet people's needs.

7.4 Quality Control

Quality control ensures that a program's activities take place as designed. Quality control activities also may uncover flaws in design and thus point to changes that could improve quality. For foods, the main objective of quality control is to ensure that all processes (from Farm to Table) were controlled. Quality control includes day-to-day supervision and monitoring to confirm that activities are proceeding as planned and staff members are following guidelines. It also includes periodic evaluations that measure progress toward program objectives. Good quality control requires that programs develop and maintain:

- Measurable indicators of quality,
- Timely data collection and analysis, and
- Effective supervision.

7.5 Quality Assurance

Quality Assurance has become an increasingly important part of the food manufacturing process for both the quality and safety of food. However many are unsure as to the exact meaning of this and how it can be used to encompass broad bands of the work already underway and used to demonstrate how effectively managed a food business is.

Quality Assurance can be defined as a series of planned actions necessary to provide the customer with the product that they expect. A lot of businesses have well developed Quality Assurance systems that have emerged over a period of time and are regularly reviewed. By documenting and recording all the steps involved it demonstrates a "control" of their product and process but also allows feedback into the production process to develop it further.

Quality Assurance is a proactive process which attempts to stop errors happening in the process allowing it to be "right first time". It involves using the HACCP to identify the areas of concern-the right control points to be evaluated at corrective actions put in place and the documentation supporting this to be recordedand kept in case of problems later on.

There are three main components in the Quality Assurance process:

1) The specification (What is to be done)
2) Documented instruction (How it is to be done)
3) Recording system (That it has been done)

Whatever the business size you are dealing with these three areas can be written and recorded to demonstrate that you have control of your product and process.

The benefits of applying Quality Assurance are very real. By ensuring that procedures are soundly based and efficient it should save money and give assurance to customer satisfaction. Making faulty product costs money and could harm customer satisfaction badly.

Methods applied in Quality Assurance systems need to be applicable for the type of process and environment they are in and work within the resources available.

To achieve the objectives of Quality Assurance it is necessary to design and plan as relevant:

1) Raw Material Specification

2) Ingredients formulation

3) Processing equipment and environment

4) Processing methods and conditions

5) Intermediate in-specifications

6) Appropriate labeling specifications

7) Specifications for quantity per pack

8) Specifications for management and control procedures

9) Specified distribution system and cycle

10) Appropriate storage, handling and distribution instructions

Quality Assurance must be seen as beneficial to the ice cream industry as a system that will provide a contribution to improving food quality and safety. But it is not an easy implication and the natural variation in raw materials due to seasonal changes needs to be considered. The concept is to achieve "zero defects" which if written correctly will be seen as a realistic target within a process and which is why the government have allowed a defense of "due diligence".

Once you have set up your Quality Assurance system it will need to be checked regularly to ensure how effective it is. Checking all the documentation to ensure it is all filled in, reviewing the process, and using the HACCP plan to ensure all the Critical Control Points have been checked are all part of this. Challenging the processes and control points will be the areas any auditor coming to your site will look at.

There are four basic tenets or principles of quality assurance. They are:

1) **Client Focus**

Products/Services should be designed so as to meet the needs and expectations of clients and community.

2) **Understanding Work as Processes and Systems**

Providers must understand the product or service system and its key processes in order to improve them.

3) **Testing Changes and Emphasizing the Use of Data**

Changes are tested in order to determine whether they yield the required improvement. Data are used to analyze processes, identify problems, and to determine whether the changes have resulted in improvement.

4) **Teamwork**

Improvement is achieved through the team approach to problem solving and quality improvement.

7.6 Quality Improvement

Quality improvement (QI) is a revolutionary idea in food quality. The idea is to raise the level of food quality—no matter how good it may already be—through a continuous search for improvement. QI asks managers, providers, and other staff members not just to meet the standards but rather to exceed them—indeed, to raise the norms.

There are many ways to improve quality, such as enforcing or revising standards, strengthening supervision, and asking managers or technical experts to redesign a process. The concept of QI, which is grounded in the quality movement in industry, usually involves a team-based problem-solving approach, however.

In QI groups of staff members at the national, district, or facility level work together to identify and resolve problems that compromise the quality of foods. They base their decision-making on data rather than assumptions, use diagnostic and analytic tools, and follow a systematic process. An individual supervisor or manager can take this same approach, but QI seeks to harness the managers and personnel at every level to improve the quality of products or services.

A growing number of food corporations have adapted the principles of team-based problem-solving. While QI initiatives in developing countries are too recent to demonstrate long-term impact, there is evidence that the team-based approach has helped individual facilities and entire programs use resources more efficiently and improve administration and service delivery.

At the same time, however, experience with QI teams has found that they are difficult to implement, they work slowly, and they may sidestep the most difficult problems. Other management techniques may improve the quality of services more efficiently and at less cost. In Indonesia, for example, staff from the Quality Assurance Project (QAP) gained the impression that about half of the observed quality deficiencies could be corrected by reinforcing standards, while another quarter required small-scale Operations Research to develop solutions that could be widely applied. Only about one-quarter needed team-based problem-solving. Quality initiatives in developing countries have relied more heavily on conventional management approaches than on QI teams.

New Words

staffing   安置职工
recruit   新兵，新分子，新会员
simultaneously   同时地
empower   授权与，使能够
diagnose   诊断
coordination   同等，调和
quality design   质量设计
mission   使命，任务
strength   优势

quality control　质量控制
flaw　缺点，裂纹，瑕疵
supervision　监督，管理
monitor　监控
quality assurance　质量保证
encompass　包围，环绕
demonstrate　示范，证明，论证
faulty　有过失的，有缺点的，不完美的，不完善的
specification　详述，规格，说明书，规范
formulation　用公式表示，工艺
client focus　以顾客为关注焦点
quality improvement　质量改进
assumption　假定，设想，担任，承当，假装，作态

Exercises

Ⅰ. Answer the following questions according to the article.

1. What is the definition of food quality management?
2. What are the functions of quality management?
3. What are the principles of quality management?
4. How to do quality design?
5. What are the principles of quality assurance?
6. What is the ideal of quality improvement?

Ⅱ. Choose a term from what we have learnt to fill in each of the following blanks. Change the word form where necessary.

1. Quality management include ＿＿＿＿, ＿＿＿＿, ＿＿＿＿ and ＿＿＿＿.
2. Creative problem solving is broader than ＿＿＿＿, ＿＿＿＿ or ＿＿＿＿.
3. It is important to ＿＿＿＿＿＿, in order to diagnose and solve organizational problems and measure progress.
4. Quality design begins with defining the organization's mission, including its ＿＿＿＿ with an eye to quality.
5. Quality control includes day-to-day ＿＿＿＿ and ＿＿＿＿ activities are proceeding as planned.
6. Improvement is achieved through the team approach to ＿＿＿＿ and ＿＿＿＿.
7. Quality Assurance can give assurance to ＿＿＿＿＿＿＿.

Unit 8　Quality Control Tools

Production environments that utilize modern quality control methods are dependant upon statistical literacy. The tools used therein are called the seven quality control tools. These include:

1. Check sheet
2. Flowchart
3. Cause and Effect Diagram
4. Control Chart
5. Histogram
6. Scatter Plot
7. Pareto Diagram

8.1 Check sheet

Check sheets help organize data by category. They show how many times each particular value occurs, and their information is increasingly helpful as more data are collected. More than 50 observations should be available to be charted for this tool to be really useful. Check sheets minimize clerical work since the operator merely adds a mark to the tally on the prepared sheet rather than writing out a figure (Table 8.1). By showing the frequency of a particular defect (e.g., produce of bread) and how often it occurs in a specific location, check sheets help operators to find spot problems. The check sheet example shows a list of defects of bread on a production line covering a week's time. One can easily see where to set priorities based on results shown on this check sheet. Assuming the production flow is the same on each day, the part with the largest number of defects carries the highest priority for correction.

Table 8.1 Quality check sheet of bread

Defective Item	9/1(M)	9/2(T)	9/3(W)	9/4(T)	9/5(F)	Total
Size	5	3	6	3	4	21
Shape	2	0	5	1	0	8
Color	4	2	3	5	0	14
Weight	1	5	0	2	1	8
Other	0	1	1	0	0	2
Total	12	11	15	11	5	45

8.2 Flow chart

A flow chart provides a visualization of a process by use of symbols that represent different types of actions, activities or situations. These symbols represent (1) start/end of the process, (2) a process or part of a process, (3) inspection, (4) decision, and (5) transport, five activities that can be used to describe a wide variety of complete processes. Other symbols can be used for different specific types of processes; for example, a description of a software program might contain specific symbols for input/output, storage, and so forth. The user of flow charts should be able to utilize any symbols that are needed to represent the processes that are being dealt with. The figure below displays a typical process flow that describes the simple process for getting a cup of coffee (Figure 8.1). Symbols used are connected with 'links' which show the flow of information between the symbols used to represent the steps in the process.

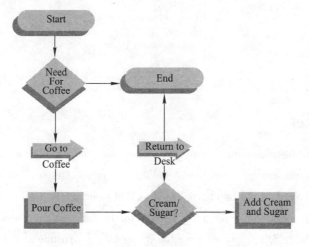

Figure 8.1 Get a cup of coffee

Steps for Creating a Flow Chart are:

1. Familiarize the participants with the flowchart symbols.

2. Brainstorm major process tasks. Ask questions such as "What really happens next in the process?", "Does a decision need to be made before the next step?", or "What approvals are required before moving on to the next task?".

3. Draw the process flowchart using the symbols on a flip chart or overhead transparency. Every process will have a start and an end (shown by elongated circles). All processes will have tasks and most will have decision points (shown by a diamond).

4. Analyze the flowchart for such items as:

Time-per-event (reducing cycle time)

Process repeats (preventing rework)

Duplication of effort (identifying and eliminating duplicated tasks)

Unnecessary tasks (eliminating tasks that are in the process for no apparent reason)

Value-added versus non-value-added tasks

8.3 Cause and Effect Diagram

The cause and effect diagram is also called the fishbone chart because of its appearance (Figure 8.2) and the Ishakowa chart after the man who popularized its use in Japan. Its most frequent use is to list the cause of particular problems. The lines coming off the core horizontal line are the main causes and the lines coming off those are sub-causes.

A fish bone diagram displays all contributing factors and their relationships to the outcome to identify areas where data should be collected and analyzed. The major areas of potential causes are shown as the main bones, e.g., materials, methods, people, measurement, machines, and design

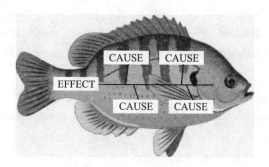

Figure 8.2 Fishbone chart

(Figure 8.3). Later, the sub-areas are depicted. Thorough analysis of each cause can eliminate causes one by one, and the most probable root cause can be selected for corrective action. Quantitative information can also be used to prioritize means for improvement, whether it be to machine, design, or operator.

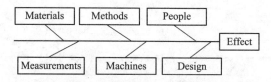

Figure 8.3 Cause and effect diagram

Stepsin Constructing a Cause and Effect Diagram

1. Prepare a flip chart or an overhead transparency of the following template:

2. Write the issue (problem or process condition) on the right side of the Cause and Effect Diagram.

3. Identify the major cause categories and write them in the four boxes on the Cause and Effect Diagram. You may summarize causes under categories such as:

Methods, Machines, Materials, People, Places, Procedures, People, Policies, Surroundings, Suppliers, System, Skills

4. Brainstorm potential causes of the problem. As possible causes are provided, decide as a group where to place them on the Cause and Effect Diagram. It is acceptable to list a possible cause under more than one major cause category.

5. Review each major cause category. Circle the most likely causes on the diagram.

6. Review the causes that are circled and ask "Why is this a cause?" Asking "why" will help get to the root cause of the problem.

7. Reach an agreement on the most probable cause(s).

8.4 Control Chart

A control chart displays statistically determined upper and lower limits drawn on either side of a process average. This chart shows if the collected data are within upper and lower limits previously determined through statistical calculations of raw data from earlier trials.

The construction of a control chart is based on statistical principles and statistical distributions, particularly the normal distribution. When used in conjunction with a manufacturing process, such charts can indicate trends and signal when a process is out of control. The center line of a control chart represents an estimate of the process mean; the upper and lower critical limits are also indicated. The process results are monitored over time and should remain within the control limits; if they do not, an investigation is conducted for the causes and corrective action taken. A control chart helps determine variability so it can be reduced as much as is economically justifiable.

In preparing a control chart, the mean upper control limit (UCL) and lower control limit (LCL) of an approved process and its data are calculated. A blank control chart with mean UCL and LCL with no data points is created; data points are added as they are statistically calculated from the raw data.

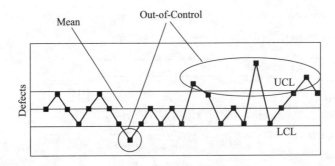

Figure 8.4　Control chart

Consider the figure above, which shows an example of a control chart. In this example, the numbers of defects are plotted against the samples taken from some manufacturing process. For each sample, the average numbers of defects found are plotted. For example, if one hundred units are selected each day for testing, the number defective would be plotted on this chart. If samples are taken at regular intervals, for example, day by day, there will be variability in the number of defects found. Note in this example chart that the variation appears random. Using statistical methods, the upper and lower bounds of the process for normal variation are calculated and these boundaries plotted on the chart as the upper and lower control limits (UCL, LCL). Some points do indeed appear on the chart above and below the control limits. The points that exceed the expected limits are said to be "out-of-control". Typically a manufacturing line manager's task is to return the process to normal variation by determining what has caused the variation that exceeded expected limits. Of course, the real problem is not maintaining level of defects but decreasing the number. Hence, these charts could be used to examine long term trends and use the charts to determine if long term improvement is taking place. Such examination could be done statistically by plotting trends over time to see if there is a consistent pattern of improvement taking place.

8.5　Histogram

The figure above shows a typical histogram. It is a bar chart showing a distribution of variable.

What distinguishes the histogram from a check sheet is that its data are grouped into rows so that the identity of individual values is lost. Commonly used to present quality improvement data, histograms work best with small amounts of data that vary considerably. When used in process capability studies, histograms can display specification limits to show what portion of the data does not meet the specifications.

After the raw data are collected, they are grouped in value and frequency and plotted in a graphical form (Figure 8.5). A histogram's shape shows the nature of the distribution of the data, as well as central tendency (average) and variability. Specification limits can be used to display the capability of the process.

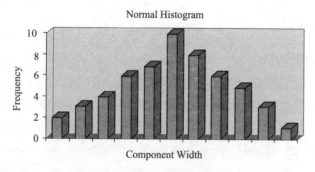

Figure 8.5　Histogram

8.6　Scatter Plot

A scatter diagram shows how two variables are related and is thus used to test for cause and effect relationships. It cannot prove that one variable causes the change in the other, only that a relationship exists and how strong it is. In a scatter diagram, the horizontal (x) axis represents the measurement values of one variable, and the vertical (y) axis represents the measurements of the second variable.

The figure below shows a plot of two variables, in this example, reactant concentration versus reactive velocity. As shown, as the reactant concentration increases, so does the reactive velocity. These variables are said to be positively correlated, that is, if one increases so does the other. The line plotted is a "regression" line which shows the average linear relationship between the variables. If the line sloped negatively, the variables would be negatively correlated, that is, when one increases, the other decreases, and vice versa. When no regression line can be plotted and the scatter plot appears to simply be a ball of diffuse points, then the relationship between the variables are said to be uncorrelated.

The utility of the scatter plot for quality assessment is the determination by measuring variables in a process to see if any two or more variables are correlated or uncorrelated. This information can be useful in the "relations" diagram discussed below. The specific utility of finding correlations is to infer causal relationships among variables and ultimately find root causes of problems. The scatter plot is simply one of the tools which can contribute additional small amounts of information to a

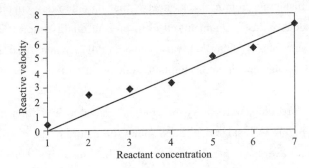

Figure 8.6 Relationship between reactant and reactive velocity

quality assessment.

8.7 Pareto Diagram

Pareto diagrams are named after Vilfredo Pareto, an Italian sociologist and economist, who invented this method of information presentation toward the end of the 19th century. The figure above shows an example of a Pareto diagram. The Pareto shows the distribution of items and arranges them from the most frequent to the least frequent with the final bar being misc. The figure above shows an example of a Pareto diagram. The chart appears much the same as a histogram or bar chart, except that the bars are arranged in decreasing order from left to right along the abscissa. The fundamental idea of use of Pareto diagrams for quality improvement is the ordering of factors that contribute to a quality function. For example, suppose that you have identified the following factors in a problem and their numerical and percentage contribution to the problem. In the right column are shown two numbers: On the left is an actual count contributing to the factor and on the right is the percentage that this number is of the total.

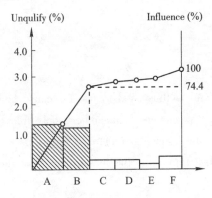

Figure 8.7 Pareto diagram

8.8 Summary

The basic quality tools provide a simple yet powerful set of tools for the visualization of information. Some of the tools fall in the category of common sense (e.g., check sheet and Pareto), some implement basic statistics (e.g., histogram, scatter plot, and stratification), another implements a widely used quality control technique (the control chart) and finally, the last tool, the fishbone diagram implements a brainstorming tool. Which of these tools are most important? That is difficult to say; however, it is possible to comment on which tools are used most often in practice. The control chart is probably most widely used of any of these tools, simply for historical reasons. Rooted in the practice of statistical process control, the use of control charts has become widely used in North American industry. The fishbone appears to be the most popular tool for use among workers that are not statistically inclined. The check sheet can be used for tabulating information, but is otherwise not a very sophisticated tool. Use of histograms, stratification, and scatter plots appears to be less prevalent, simply because the average worker using the tools does not have the necessary statistical background to interpret the results of collecting and analyzing data in this way. This is particularly true of the stratification chart, which requires making analytical judgements and decisions about how to segment data being collected, a task that is not easy for the average worker on an industrial team.

In sum, the basic tools provide a robust set of tools which permit viewing problems from different perspectives. The problem with the tools is the lack of use of the tools among workers. They cannot solve their own problems without understanding the way the tools operate and how the tools can assist them in understanding their own problems.

New Words

quality control tool 质量控制工具
check sheet 检查表
flowchart 流程图
Cause and Effect Diagram = fish bone diagram 因果图,鱼骨图
control chart 控制图
histogram 柱状图
scatter Plot 散布图
Pareto diagram 柏拉图
statistical 统计的,统计学的
visualization 使看得见的,清楚地呈现在心
brainstorm 头脑风暴法
flip 无礼的,冒失的,轻率的
transparency 透明,透明度
diamond 菱形
eliminate 排除,消除
duplicate 复制的,副的,两重的,两倍的
come off 离开,举行,实现,成为
depict 描述,描写
prioritize 把……区分优先次序
trial 试验,考验
conjunction 联合,关联
normal distribution 正态分布,正常分布
upper control limit (UCL) 上控制线
lower control limit (LCL) 下控制线
variation 变更,变化

variable　【数】变数，可变物，变量
graphical　绘成图画似的，绘画的
horizontal (x) axis　横轴(x轴)
vertical (y) axis　纵轴(y轴)
reactant　反应物
velocity　速度，速率，迅速，周转率
regression　衰退
Vilfredo Pareto　维尔弗雷多·帕累托，意大利社会学和经济学家
misc　各色各样混在一起，混杂的，多才多艺的

Exercises

Ⅰ. Answer the following questions according to the article.

1. How many quality control tools do you use in your work?
2. What is the aim to list a check sheet?
3. How to draw a flow chart?
4. What is the definition of cause and effect diagram?
5. How to analyze the major cause categories by cause and effect diagram?
6. How to judge whether a process to be controlled or not by control chart?
7. How to draw a histogram?
8. What are the roles of Pareto diagram?

Ⅱ. Choose a term from what we have learnt to fill in each of the following blanks. Change the word form where necessary.

1. Modern quality control methods are dependant upon _____.
2. By showing the _____ of a particular defect and how often it occurs in a specific location, check sheets help operators to find _____.
3. Start/end of the process, a process or part of a process, _____, _____, _____ can be used to describe a wide variety of complete processes in flow chart.
4. Cause and effect diagram include the _____ causes and the _____ causes.
5. The construction of a control chart is based on _____ and _____, particularly the normal distribution.
6. Histogram is a bar chart showing a _____.
7. A scatter diagram shows how two variables are related and is thus used to test for _____.
8. Pareto diagram was invented by _____, who is an Italian sociologist and economist.

Unit 9　Crisis Management for Food Industry

9.1　Crisis Definitions

A crisis is defined by the dictionary as a "critical moment or turning point". A business book, on the other hand, might define a crisis as a substantial, unforeseen circumstance that can potentially jeopardize a company's employees, customers, products, services, fiscal situation, or reputation. Both definitions contain an element of urgency that requires immediate decisions and actions from people involved.

The definition given by the American Institute for Crisis Management (ICM) for the word "crisis" underscores the association of a crisis with media coverage. ICM defines "crisis" as "*a significant business disruption which stimulates extensive news media coverage. The resulting public scrutiny will affect the organization's normal operations and also could have a political, legal, financial, and governmental impact on its business*".

The basic causes of a business crisis are four in number:

1. Acts of God (storms, earthquakes, volcanic action, etc.)
2. Mechanical problems (ruptured pipes, metal fatigue, etc.)
3. Human errors (the wrong valve was opened, miscommunication about what to do, etc.)
4. Management decisions/indecision (the problem is not serious, nobody will find out)

Most of the crises ICM has studied fall in the last category and are the result of management not taking action when they were informed about a problem that eventually would grow into a crisis. Crisis events generally fall into two basic types based on the amount of warning time.

9.1.1　Sudden Crisis

A sudden crisis is defined as: A disruption in the company's business which occurs without warning and is likely to generate news coverage and may adversely impact:

1. Our employees, investors, customers, suppliers or other publics
2. Our offices, franchises or other business assets
3. Our revenues, net income, stock price, etc.
4. Our reputation and ultimately the good will listed as an asset on our balance sheet

A sudden crisis may be:

a. A business-related accident resulting in significant property damage that will disrupt normal business operations

b. The death or serious illness or injury of management, employees, contractors, customers, visitors, etc. as the result of a business-related accident

c. The sudden death or incapacitation of a key executive

d. Discharge of hazardous chemicals or other materials into the environment

e. Accidents that cause the disruption of telephone or utility service

f. Significant reduction in utilities or vital services needed to conduct business

g. Any natural disaster that disrupts operations, endangers employees

h. Unexpected job action or labor disruption

i. Workplace violence involving employees/family members or customers

9.1.2　Smoldering Crisis

A smoldering crisis is defined as: Any serious business problem that is not generally known inside or outside the company, which may generate negative news coverage if or when it goes "public" and could result in more than a predetermined amount in fines, penalties, legal damage awards, unbudgeted expenses and other costs. Examples of the types of smoldering business crises that would prompt a call to the Crisis Management Team would include:

a. Sting operation by a news organization or government agency

b. OSHA or EPA violations which could result in fines or legal action

c. Customer allegations of overcharging or other improper conduct

d. Investigation by a federal, state or local government agency

e. Action by a disgruntled employee such as serious threats or whistle-blowing

f. Indications of significant legal/judicial/regulatory action against the business

g. Discovery of serious internal problems that will have to be disclosed to employees, investors, customers, vendors and/or government officials

In some instances crisis situations may be either sudden or smoldering, depending on the amount of advance notice and the chain of events in the crisis. Examples would include:

Adverse government actions	Computer tampering
Anonymous accusations	Damaging rumors
Competitive misinformation	Discrimination accusations
Confidential information disclosed	Equipment, product or service sabotage
Misuse of chemical products	Industrial espionage
Disgruntled employee threats	Investigative reporter contact
Employee death or serious injury	Judicial action
Employee involved in a scandal	Labor problems
Licensing disputes with local officials	Lawsuit likely to be publicized
Extortion threat	Security leak or problem
False accusations	Severe weather impact on business
Incorrect installation of equipment	Sexual harassment allegation
Grand jury indictment	Special interest group attack
Grass roots demonstrations	Strike, job action or work stoppage
Illegal actions by an employee	Terrorism threat or action
Indictment of an employee	Illegal or unethical behavior of an employee
Major equipment malfunction	Union organizing actions
Nearby neighbor, business protest	Whistleblower threat or actions

Many empirical researches into business crisis events indicate that most sudden crises also generate "aftershocks" in the form of smoldering crises which occur as the government, media and internal investigations into the cause of the crisis uncover specific problems that were not know previously. Many of those aftershocks are included in the list directly above.

9.2 Crisis Management

Crisis Management is the process of preparing for and responding to an unpredictable negative event to prevent it from escalating into an even bigger problem, or worse, exploding into a full-blown, widespread, life-threatening disaster. Crisis management involves the execution of well-coordinated actions to control the damage and preserve or restore public confidence in the system under crisis.

In the context of corporate governance, excellent crisis management is a "must" whenever a crisis occurs because of the crisis' enormous potential impact on the company's reputation and financial standing. Poor handling of a crisis situation can ruin the confidence of the customers or the public in a company and jeopardize its survival, a situation that normally takes a long time to correct, if it still is reparable at all. Such is the importance of public perception of a company's handling of a crisis situation that media coverage management has become an important ingredient of crisis management.

Crisis management doesn't start only when a crisis arises and ends when "the last fire has been put out". Crisis management requires actions before a crisis happens, while the crisis is unfolding, and after the crisis has ended. In fact, crisis management is divided into these three stages:

1) Pre-incident stage, which involves identification of potential crisis situations and developing contingency plans for responding to each of them;

2) Incident stage, which involves management of an ongoing actual crisis situation itself; and

3) Post-incident stage, which includes corrective and preventive actions to preclude the recurrence of the same crisis situation and business recovery actions to restore public confidence in the brand or the company.

There are many different ideas or theories on how to best manage a crisis situation. These differing ideas, nonetheless, have some common elements:

1) The need to anticipate potential crisis situations and prepare for them;

2) The need to provide accurate information during a crisis;

3) The need to react as quickly as possible to the situation;

4) The need for a response that comes from the top; and

5) The need for long-term solutions.

Anticipating potential crisis situations that a company may encounter and formulating and documenting contingency action plans for them are a basic requirement of any crisis management program. These plans should also be well-rehearsed by all employees, so the conduct of regular drills is also needed. Any company must be prepared to deal with fires, bomb threats, personnel violence, and natural disasters such as earthquakes and tornadoes. In the semiconductor industry, the discovery of a life-threatening device reliability issue, process gas leaks, hazardous chemical spills, and even the sudden loss of a major supplier are examples of crisis situations.

One of the hardest things to contend with during a crisis is the craving of customers or the public for a constant supply of information. Accurate information pertaining to a crisis is often not readily available at once, and is too bitter to announce to the public once they become available. If the crisis does not concern the public, then a company may get away with staying quiet about it.

Otherwise, the press will usually be all over the place within hours. In such a case, there is no choice—most experts agree that it is better to provide accurate information, no matter how painful they sound, than to manipulate the situation by giving false information to the public, which often backfires with tremendous repercussions. Thus, a company that has developed a culture of internal secrecy and manipulation is of great disadvantage in this respect, because they would find it difficult to provide honest information and subsequently resort to the thing they usually do: hide the truth. Public Relation work for high-profile crises is something that any management team must be well-trained for.

Being indecisive, playing time or inability to get accurate information quickly can be disastrous during a crisis. Management must act swiftly and decisively to contain the problem, assess affected goods, ensure business continuity, allay public fears, and preserve company reputation even while the crisis is still unfolding. Since a crisis by definition is unpredictable, a company needs to have a system for assembling a crisis management team that knows what to do within an hour after a crisis occurs, 24 hours, all day, and all year long.

The visibility of top management people during a crisis is highly recommended by experts because it assures the public that the problem is getting due corporate attention. Management must also actively pursue long-term corrective and preventive actions to avoid being in the same crisis situation again.

A comprehensive crisis management program includes the following components:

1) An emergency response, which consists of all activities pertaining to safe management of immediate physical, health, and environmental effects of the crisis;

2) Business continuity, or the company's ability to continue delivering goods and services despite the crisis;

3) Crisis communications, which pertains to the internal and external PR management activities during a crisis;

4) Humanitarian assistance, or the company's efforts to alleviate the physical, emotional, and psychological effects of the crisis on other people; and

5) Drills and exercises that allow personnel to rehearse what they need to do in a crisis situation.

9.3 International Crisis Management

International crisis management takes many forms. Within the food industry, the primary focus involves product quality issues related to product contamination, mislabeling, and product adulteration, which may result in the withdrawal/recall of products from the market place. The movement of food products between countries greatly increases the complexity.

Once a crisis situation is identified, it is critical to have a comprehensive crisis management plan to guide your company through the event. All functional groups within the company will be involved in the investigation, execution, or public relations of a recall. It is essential that each group and individual knows their role and has the resources available to accomplish their required tasks. Outside third party resources will often be required to support your efforts: public relations, media contacts, legal counsel, government liaisons, testing laboratories, university personnel, retrieval services, and crisis consulting services.

The crisis management plan must be an "evergreen" document, which is continuously maintained and updated to address the changing regulatory environment as well as internal personnel changes. "Recall Simulations" should be a part of the plan to regularly practice execution of the plan and to train personnel in real life situations. The regulatory requirements and the consumer environment vary greatly in different world areas. TheU. S. plan will always need to be adapted for local conditions.

New Words

jeopardize　危害
fiscal　财政的, 会计的
reputation　名誉, 名声
score　划线于……下, 强调; 底线
scrutiny　详细审查
earthquake　地震
volcanic　火山的
rupture　破裂, 裂开, 断绝(关系等), 割裂
fatigue　疲乏, 疲劳, 累活
valve　阀
indecision　优柔寡断
sudden crisis　突发危机
asset　资产, 有用的东西
contractor　订约人, 承包人
endanger　危及
smolder　郁积, 闷烧; 闷烧
smoldering crisis　缓慢危机
OSHA = Occupational Safety and Health Act　职业安全与卫生条例【美】
EPA abbr. Environmental Protection Agency　美国环保署
whistleblowing　告密, 揭发
aftershock　余震
perception　感知, 感觉

contingency　偶然，可能性，意外事故，可能发生的附带事件
rehearse　预演，排演，使排练，复述，练习
tornado　旋风，龙卷风，大雷雨，具有巨大破坏性的人（或事物）
craving　渴望
pertaining　与……有关系的，附属……的，为……固有的(to)
manipulate　（熟练地）操作，使用（机器等），操纵（人或市价、市场）
backfire　逆火，产生反效果，事与愿违
repercussion　弹回，反响，（光、声等的）反射
swiftly　很快地，即刻
allay　减轻，减少
unfold　打开，显露，开展，阐明
visibility　可见度，可见性，显著，明显度，能见度
humanitarian　人道主义者
liaison　联络，（语音）连音
retrieval　取回，恢复，修补，重获，挽救
recall simulations　模拟召回

Exercises

Ⅰ. Answer the following questions according to the article.

1. What is the definition of crisis? How to comprehend it's meaning for yourself?
2. What is the definition of sudden crisis? Please give several examples of sudden crisis.
3. What are basic causes of a business crisis?
4. What is the definition of smoldering crisis? Please give several examples of smoldering crisis.
5. What is the definition of crisis management? How to execute crisis management?
6. What are common elements of crisis management?
7. How many components had been included a comprehensive crisis management program?
8. How to implement a crisis management for food industry?

Ⅱ. Choose a term from what we have learnt to fill in each of the following blanks. Change the word form where necessary.

1. A crisis can potentially _____ a company's employees, customers, products, services, fiscal situation, or _____.

2. Earthquakes are acts of _____, metal fatigue is a _____ problem and miscommunication is a human error.

3. Crisis events generally fall into two basic types based on the amount of warning time, they are _____ and _____.

4. Unexpected job action or labor disruption is a _____ crisis, investigation by a federal, state or local government agency is a _____ crisis.

5. Most sudden crises generate "_____" in the form of smoldering crises.

6. Crisis Management is the process of preparing for and responding to an _____ event.

7. Crisis management is divided into three stages： _____, _____ and _____.

8. Within the food industry, the primary focus involves product quality issues related to product contamination, _____ and _____.

Unit 10　Foods Entry-Exit Inspection and Quarantine

10.1　Introduction to Foods Entry-Exit Inspection and Quarantine

10.1.1　Inspection

Inspection is defined as a formal or official examination. Food inspection includes sensory, physical, chemical and biological examination.

10.1.2　Quarantine

Quarantine is defined as a period of time during which a vehicle, person, or material suspected of carrying a contagious disease is detained at a port of entry under enforced isolation to prevent disease from entering a country.

The word "quarantine" derives from the Italian *quaranta dei* (forty days), a reminder that the custom of segregating putatively infected persons, and the ships on which they were traveling, originated in the maritime empire of Venice in the fourteenth century. The length of time probably relates to the biblical story of the forty days Jesus spent in the wilderness, not to any real knowledge of the mode of transmission of infection. The rise of the practice, however, suggests that there was some understanding of the concept of contagion even if there was no empirical knowledge of infective periods and incubation times of the plagues that afflicted medieval Europe. Few infectious diseases have an incubation time or infective period greater than forty days. An exception is rabies, which may not declare itself for many months. That is why animals that may have been exposed to rabies are quarantined for many months when they arrive in countries where rabies does not exist.

Animal and plant quarantine procedures are often more important than human quarantine now that many of the most dangerous contagious diseases can be kept under observation without such draconian restrictions as formerly required. The economic importance of agriculture and animal husbandry in many countries makes it absolutely essential to exclude diseases that might wipe out valuable cattle herds or destroy a season's harvest.

10.2 China's Entry-Exit Commodity Inspection and Quarantine System

10.2.1 Inspection and Quarantine Authorities

In China, the department in charge of the inspection and quarantine of commodities is the State Administration for Entry-Exit Inspection and Quarantine of the People's Republic of China (SAIQ), which was established on March 29, 1998. It was created by a merger of three government departments, namely, the State Administration of Import and Export Commodity Inspection, the State Animal and Plant Quarantine Administration under the Ministry of Agriculture, and the National Health and Quarantine Administration under the Ministry of Health.

SAIQ is the administrative institution enforcing the law in the field of entry/exit commodity health inspection and quarantine, animal and plant quarantine and the inspection, survey, and supervision of commodities.

Apart from SAIQ, another quarantine office is China National Import & Export Commodities Inspection Corporation (CCIC). It is a comprehensive inspection and survey agency duly authorized by the Chinese Government. Founded in 1980, it is the first national non-governmental organization of its kind inChina, focusing its principal activities on the field of import and export commodity inspection and certificate inspection.

10.2.2 Main Inspection and Quarantine Operation

Import and export commodity inspection covers the inspection of a commodity's quality, quantity, weight, safety and sanitation. All commodities listed in the Catalogue of Import and Export Commodities subject to Statutory Inspection, or commodities subject to statutory inspection otherwise specified by laws and regulations, must be subject to inspection by the Chinese entry/exit inspection and quarantine authorities or other designated institutions. The consignee of the imports subject to statutory inspection must register with the inspection and quarantine authorities at the port of discharge; and the consignor of the imports and exports subject to statutory inspection shall declare commodity inspection within the period and at the place specified by the inspection and quarantine authorities. Commodities subject to statutory inspection may not be marketed or used before being inspected, and export commodities may not be exported if found inferior in quality through inspection.

All food for export or import must be inspected and up to the required standards. Otherwise, they may not be imported or exported.

All animals, plants, and their products for import, export and in transit, as well as materials used for carrying animals and plants, shall be subject to quarantine and inspection. Any person entering or leaving the country shall be placed in quarantine by the inspection and quarantine authorities to discover any possible communicable diseases, and the necessary prophylaxis and control measures shall be adopted by the authorities. Theyshall also inspect the sanitary conditions at border ports and on the means of transport inbound and outbound at ports.

10.2.3　Functions of the Inspection and Quarantine Statement and Certificate

The inspection and quarantine statement on commodities to be imported and exported is to certify the quality of the commodities, and the certificate of qualification can be useful at least for three purposes, as follows:

1. For statutory inspected commodities, the quarantine certificate of qualification is one of the necessary documents for their entry or exit of a country. Customhouses will give clearance for them upon receiving the certificate.

2. For those commodities to be inspected and quarantined as specified in a business agreement signed between two parties, the qualified certificate of commodities is used to certify the quality of commodities up to the standard specified in the agreement. Both parties to the agreement usually at the same time agree on their appointed institution, place and time for inspection and quarantine in the terms. Both the parties can claim for any loss and launch a suit, if necessary, only under the condition that they have already strictly executed the terms on inspection and quarantine as specified in the agreement and acquired a qualified certificate of quarantine.

3. Even if the terms on quarantine are not included in an agreement, they may have their commodities quarantined voluntarily. For the consignor or exporter, a qualified certificate of commodities inspected before loading on board ship can certify the quality of commodities up to the standard specified in the agreement when they are discharged. In case of a problem of quality arising when they have reached the port of termination, the consignor or exporter can use the said qualified certificate to reject the recourse of the consignee for losses. The quality problem may be caused in transit by the carrier's negligence or by improper stowage; for the consignee or importer, inspecting commodities immediately after they reach the port of termination can ensure the quality condition of the commodities. If a quality problem arises, timely inspection before they are transferred and treated will provide a convincing evidence for the consignee to claim for losses. Therefore, inspection and quarantine are equally beneficial to both sides in a trade.

10.3　Supervision and Inspection of Foodstuffs

10.3.1　Inspection of Imported Food

All imported food (including beverages, liquor and sugar), food additives, food containers, packaging materials, and food utensils and equipment must be declared to the inspection and quarantine authorities for health supervision and inspection. Under the management of inspection and quarantine authorities, imported food is classified into different categories by hazard ratings and inspected according to national health standards. Only those foodstuffs that meet requirements can enter China.

10.3.2　Inspection of Exported Food

All food for export (including finished products and raw materials for human consumption and food prepared according to traditional methods with medicinal ingredients) must undergo inspection. Foodstuffs not inspected or not up to standard may not be exported.

10.3.3 Registration of Imported Food and Animal and Plant

CIQ is responsible for formulating, revising and publishing the *Catalogue of Imported and Exported Food and Animal and Plant Products Subject to Health Registration*. The products in the catalogue mainly include processed food that can be consumed directly, such as canned food, beverages, liquor and condiments, as well as semi-finished products and raw materials, such as meat, aquatic products and vegetables.

Foreign food manufacturers may apply for registration with CIQ through the local food and hygiene authorities at their resident country. Those that meet CIQ requirements will be added to CIQ's list of countries and enterprises permitted to export food to China and issued special quarantine and health registration codes.

10.3.4 Registration of Exported Food and Animal and Plant

All enterprises in China engaged in the processing and storage of food for export as well as the slaughtering of animals and poultry for export must first obtain a health license from the local health department and then apply to the inspection and quarantine authorities for a registration certificate. Enterprises not granted the registration certificate may not process, produce or store food for export. Where overseas registration or recognition is required, the enterprises must apply to CIQ which is responsible for the unified handling of all the necessary foreign-related procedures. Enterprises failing to obtain approval or recognition from the importing country may not export food to that country.

10.4 Status and Prospect of China's Food Imports & Exports Safety

10.4.1 Status of China's Food Imports & Exports Safety

Food safety is one of concerns worldwide. Food-borne diseases, aroused by eating unsafe food, cause hundred millions of cases and millions of deaths. Food safety, therefore, has become ever significant in public health in the globe.

To satisfy the needs of trend development and to ensure import and export food and cosmetic safety, State department authorizes General Administration of Quality Supervision, Inspection and Quarantine to set up Import and Export Food Safety Bureau which is responsible for the administration of import and export food and cosmetic safety nationwide. The main responsibilities are as:

1. Research and regulate regulations of quality and safety supervision, inspection and quarantine of import and export food and cosmetic.

2. Organize and implement the inspection, quarantine, supervision of import and export food and cosmetic.

3. Organize and implement relevant food safety risk analysis and assessment, emergency and prevention measures.

4. Investigate significant import and export food safety and quality accidents.

The establishment of Import and Export Food Safety Bureau indicates that China attaches importance to import and export food and cosmetic safety, as well as enters into a new phase.

In order to ensure entry-exit foods safety, we must establish and perfect laws and regulations related foods. Wholesome law system is the basis of import and export food safety administration. Presently, main criterions of import and export food safety administration are as "*Food Sanitation Law*", "*Import and Export Animal and Plant Quarantine Law*", "*Import and Export Commodity Inspection Law*", "*Product Quality Law*", "*Animal Disease Prevention Law*", and so on. More than 100 supplement regulations are under execution. Import and Export Food Safety Bureau has issued 991 national standards relative to food, other sections issuing more than 1,100 industry standards relative to food processing, either. Meanwhile, General Administration of Quality Supervision, Inspection and Quarantine expedites the stipulation and revise of food quality and safety, and sanitary standards. Up to now, over 986 standards on food sanitation and its inspection methodology, food quality and its inspection methodology, food additive, food packaging, food storing and transportation, food label, etc., most of which accord with international standards.

In 2001, General Administration of Quality Supervision, Inspection and Quarantine of the People's Republic of China (AQSIQ) came into being, which is in charge of national quality, metrology, entry-exit commodity inspection, entry-exit health quarantine, entry-exit animal and plant quarantine, import-export food safety, certification and accreditation, standardization, as well as administrative law-enforcement. In terms of inspection and quarantine of import and export food, there are 35 inspection and quarantine branches directly under General Administration of Quality Supervision, Inspection and Quarantine, 280 sub-branches under inspection and quarantine branches, 163 inspection and quarantine technology centers, over 300 labs involved in food inspection, 2 institutes involved in food inspection and analysis, China Import and Export Commodity Inspection Technology Institute, and Animal and Plant Quarantine Lab, General Administration of Quality Supervision, Inspection and Quarantine. Most of these labs are equipped with more than 10,000 sets of advanced inspection instruments as gas chromatogram, liquid chromatogram, mass spectrum and atom absorb spectrum, etc., over 6,000 professionals engaged. Recently, BSE inspection lab and 23 genetically modified product inspection labs are established to meet the needs of trend development, which are versed in international advanced inspection methodology and apply successfully the methodology to food inspection and quarantine in China.

10.4.2 Problems and Prospect of China's Food Imports & Exports Safety

Asthe largest developing country in the world, with huge population and imbalanced regional development, China falls behind developed countries in agro-food safety. Some issues still remain. In the first place, food contamination incidents frequently occur. It is high time to further improve the use and administration of pesticides, veterinary drugs, chemical fertilizers, etc.. Secondly, there still remains gap in international standard adoption rate. There are around 8000 CAC standards on such toxic and harmful substances as pesticide and veterinary drug in food. Some developed countries, such as the US, EU, Japan, etc., has thousands of standards, with international standard adoption rate above 80%. China still falls behind in terms of food safety and sanitation and industry standards. In the third place, with the development of economy and trade, countries putting forward higher requirements on food safety and sanitation, more inspection items, higher inspection technology indexes and standards, it is necessary to put in more investment.

In one word, people will be always focused on food safety. With vital responsibility to administrate food safety, we must perform standardization and control from production origin, which is most reliable and efficient. To ensure food safety, and to further ensure national economy safety and international trade development, we must further improve origin inspection and control, enhance food safety administration in all stages, increase financial investment, extend technology cooperation domestic and abroad, further establish and improve food safety regulations and administration system.

New Words

inspection 检验,检查,视察
quarantine 检疫,隔离
detain 拘留,留住,阻止
putatively 推定地
maritime 海上的,海事的
contagion 传染,传染病,蔓延
incubation 孵蛋,抱蛋,熟虑
rabies 【医】狂犬病,恐水病
husbandry 管理
herd 兽群,牧群
catalogue 目录
statutory 法令的,法定的
customhouse 海关
consignor 委托者,发货人,寄件人,交付人
status 身份,地位,情形,状况
cosmetic 化妆品
expedite 加速,派出;畅通的,迅速的
stipulation 约定,约束,契约
methodology 方法学,方法论
versed 精通的

Exercises

Ⅰ. Answer the following questions according to the article.

1. What is the definition of food inspection?

2. Please tell us the origin of the word "Quarantine".

3. How/What to do in entry-exit animals, plants and their products?

4. What are the functions of inspection and quarantine statement and certificate?

5. Please write out five entry-exit laws related foods.

Ⅱ. Choose a term from what we have learnt to fill in each of the following blanks. Change the word form where necessary.

1. Food inspection includes _____, _____ and _____ examination.

2. The word "quarantine" derives from the Italian _____, which means _____.

3. SAIQ is the abbreviation of _____, AQSIQ is _____.

4. In China, authorities in charge of the inspection and quarantine of commodities include _____ and _____.

5. _____ is responsible for the administration of import and export food and cosmetic safety nationwide.

Chapter 2   Food Chemistry and Nutrition

  Knowledge of the constituents of foods and their properties is central to food science. The advanced student of food science, grounded in the basic disciplines of organic chemistry, physical chemistry, and biochemistry, can visualize the properties and reaction between food constituents on a molecular basis. The beginning student is not yet so equipped. This chapter, therefore, will be more concerned with some of the general properties of important food constituents, and how these underlie practices of food science and technology.

  Food is composed of three main groups of constituents, carbohydrates, proteins, and fats, and derivatives of these. In addition, there is a group of inorganic mineral components, and a diverse group of organic substances as the vitamins, enzymes, emulsifiers, acids, oxidants, antioxidants, pigments, and flavors. There is also the ever-present and very important constituent, water. These are so arranged in different foods as to give the foods their structure, texture, flavor, color, and nutritive value. In some instances foods also contain substances that can be toxic if consumed in large amounts.

  The above constituents occur in foods naturally. Sometimes we are not satisfied with the structure, texture, flavor, color, and nutritive value, or keeping quality of foods, and so we add other materials to foods. These may be natural or synthetic.

Unit 1   Water

  Water is abundant in all living things and, consequently, in almost all foods, unless steps have been taken to remove it. Most natural foods contain water up to 70% of their weight or greater unless they are dehydrated, and fruits and vegetables contain water up to 95% or greater. It is essential for life, even though it contributes no calories to the diet. Water also greatly affects the texture of foods, as can be seen when comparing grapes and raisins (dried grapes), or fresh and wilted lettuce. It gives crisp texture or turgor to fruits and vegetables, and it also affects perception of the tenderness of meat. For some food products, such as potato chips, salts, or sugar, lack of water is an important aspect of their quality, and keeping water out of such foods is important to maintain quality.

  Almost all food processing techniques involve the use of water or modification of water in some form: freezing, drying, emulsification, breadmaking, thickening of starch, and making pectin gels are a few examples. Further, because bacteria cannot grow without water, the water content has a significant effect on maintaining quality of the food. This explains why freezing, dehydration, or concentration of foods increases shelf life and inhibits bacterial growth.

  Water is important as solvent or dispersing medium, dissolving small molecules to form true solutions and dispersing larger molecules to form colloidal solutions. Acid and bases ionize in water; water is also necessary for many enzyme-catalyzed and chemical reactions to occur, including hydrolysis of compounds such as sugars. It is also important as a heating and cooling medium and as a cleansing agent.

Because water has so many functions that are important to a food scientist, it is important to be familiar with some of its unique properties. When modifying the water content of a food, it is necessary to understand these functions in order to predict the changes that are likely to occur during processing of such foods.

The chemical formula for water is H_2O. Water contains strong covalent bonds that hold the two hydrogen atoms and one oxygen atom together. The oxygen can be regarded to be at the center of a tetrahedron, with a bond angle of 105° between the two hydrogen atoms in liquid water and a larger angle of 109°6′ between the hydrogens in ice.

The bonds between oxygen and each hydrogen are polar bonds, having a 40% partial ionic character. This means that the outer shell electrons are unequally shared between the oxygen and hydrogen atoms, the oxygen atom attracting them more strongly than each hydrogen atom. As a result, each hydrogen atom is slightly positively charged and each oxygen atom is slightly negatively charged. Therefore they are able to form hydrogen bonds.

A hydrogen bond is a weak bond relatively to other types of chemical bonds such as covalent or ionic bonds, but it is very important because it usually occurs in large numbers and, therefore, has a significant cumulative effect on the properties of the substances in which it is found. Water can form up to four hydrogen bonds (oxygen can form hydrogen bond with two hydrogen atoms).

Water would to be expected to be gas at room temperature if compared with similar compounds in terms of their positions in the periodic table, but because of the many hydrogen bonds it contains, it is liquid. Hydrogen bonds between hydrogen and oxygen are common, not just between water molecules, but also between many other types of molecules that are important in foods, such as sugars, starches, pectins, and proteins.

Ice is less dense than water because the molecules have a smaller coordination number and cannot be packed together as tightly as water. As water freezes, its density decreases and its volume increases by about 9%. This is very significant when freezing foods with high water content. Containers and equipment must be designed to accommodate the volume increase when the product freezes, for example, mold for popsicles must allow room for expansion. This volume increase also contributes to the damage to the structure of soft fruits on freezing. As water is heated above 4℃, the increase in the average distance between molecules causes a slight decrease in density.

When ice is subjected to vacuum and then heated, it is converted into vapor without going through the liquid phase. This phenomenon is known as sublimation, and is the basis for the food processing method known as freezing drying. Coffee is an example of a food product that is freeze-dried. The process is expensive and is only used for foods that can be sold at a high price, such as coffee. The coffee beans are frozen and then subjected to a high vacuum, after which radiant heat is applied until almost all of water is removed by sublimation.

Water that can be extracted easily from foods by squeezing or cutting or pressing is known as free water, whereas water that cannot be extracted easily is termed as bound water. Bound water is usually defined in terms of the ways it is measured; different methods of measurement give different values for bound water in a particular food. Many food constituents can bind or hold onto water molecules, such that they cannot be easily removed and they do not behave like liquid water. Some characteristics of bound water include:

—It is not free to act as a solvent for salts and sugars.
—It can be frozen only at very low temperatures (below the freezing point of water).
—It exhibits essentially no vapor pressure.
—Its density is greater than that of water.

Water molecules bind to polar groups or ionic sites on molecules such as starches, pectins, and proteins. Water closest to these molecules is held most firmly, and the subsequent water layers are held less firmly and are less ordered, until finally the structure of free water prevails.

Water may also be entrapped in foods such as pectin gel, fruits, vegetables, and so on. Entrapped water is immobilized in capillaries or cells, but if released during cutting or damage, it flows freely. Entrapped water has properties of free water and no properties of bound water.

Freshness of any product is evaluated in part by the presence of water. Food items appear more wilted when free water is increasingly lost through dehydration.

Water activity, or A_w, is a ratio of the vapor pressure of water in a solution (p_s) to the vapor pressure of pure water (p_w). A_w must be high as living tissues require sufficient level of water to maintain turgor. However, microorganisms such as bacteria, mold, and yeast multiply at a high A_w. Because their growth must be controlled, preservation techniques against spoilage due to these microorganisms take into account the water activity of the food. Less bacterial growth occurs if the water level is lowered to less than 0.85 (FDA Model Food Code). Of course, there are other factors in addition to the water that must be present for bacterial growth to occur (food, optimum pH, etc.).

Jams, Jellies, and preserves are prepared using high concentrations of sugar and brines, which contain high concentrations of salt that are used to preserve hams. Sugar and salt are both effective preservatives as they lower down A_w. Salt lowers down even more effectively than sugar due to its chemical structure that ionizes and attracts water. Foods are also dehydrated or frozen to reduce the available water. Drying or freezing are common food preservation techniques.

The control of water level in foods is an important aspect of food quality as water content affects the shelf life of food. For example, foods may be more desirable either crispy or dry. Freezing and drying are common food preservation processes that are used to extend the shelf life foods because they render water unavailable for pathogenic or spoilage bacteria. If the water in foods is frozen quickly, there is less damage to the food at the cellular level. Preservatives may be added to a formulation to prevent mold or yeast growth. Humectants, which have an affinity for water, are added to retain moisture in foods.

New Words

turgor   充实,肿胀

emulsification 乳化，乳化作用
thickening 增稠
dehydration 脱水，脱水作用
colloidal 胶体的，胶质的
formula 分子式
covalent bonds 共价键
tetrahedron 四面体
positively charged 带正电荷的
negatively charged 带负电荷的
hydrogen bonds 氢键
coordination number 配位数
popsicles 冰棍，冰棒
freezing drying 冷冻干燥
free water 自由水
bound water 结合水
polar group 极性基团
entrap 截流
capillary 毛细管，毛细管的
jam 果酱
jelly 果冻
preserves 蜜饯水果
brine 盐水
pathogenic 病原的，致病的
humectant 湿润剂，湿润的

Unit 2 Carbohydrates

Among the most important types of carbohydrates are the sugars, dextrins, starches, celluloses, hemicelluloses, pectins, and certain gums. Chemically, carbohydrates contain only the elements carbon, hydrogen, and oxygen. One of the simplest carbohydrates is the six-carbon sugar glucose. Glucose and other simple sugars form ring structures of the following form:

α-D-glucose   α-D-mannose   α-D-glacatose

These simple sugars each contain six carbon atoms, twelve hydrogen atoms, and six oxygen atoms. They differ in the positions of oxygen and hydrogen around the ring. These differences in the arrangement of the elements result in differences in the solubility, sweetness, rates of fermentation by microorganisms, and other properties of these sugars.

Two glucose units may be linked together with the splitting out of a molecule of water. The result is the formation of a molecule of a disaccharide, in this case maltose:

Maltose

Common disaccharides formed in similar fashion are sucrose or cane sugar from glucose and fructose (a five-membered ring), maltose or malt sugar from two molecules of glucose, and lactose or milk sugar from glucose and galactose. These disaccharides also differ from one another in solubility, sweetness, susceptibility to fermentation, and other properties.

A larger number of glucose units may be linked together in polymer fashion to form polysaccharides. One such polysaccharide isamylose, an important component of plant starches. A chain of glucose units linked together in a slightly different way forms cellulose.

Thus the simple sugar are the building blocks of the more complex polysaccharides, the disaccharides and trisaccharides, the dextrins, which are intermediate in chain length, on up to the starches, celluloses, and hemicelluloses; molecules of these latter substances may contain several hundred or more simple sugar units. Chemical derivatives of the simple sugars linked together in long chains likewise yield the pectins and carbohydrate gums.

The disaccharides, dextrins, starches, celluloses, hemicelluloses, pectins, and carbohydrate gums are composed of simple sugars, or their derivatives. Therefore, they can be broken down or hydrolyzed into smaller units, including their simple sugars. Such breakdown in the case of amylase, a straight chain fraction of starch, or amylopectin, a branched chain fraction, yields dextrins of varying intermediate chain length, the disaccharide maltose, and the monosaccharide glucose. This breakdown or digestion can be accomplished with acid or by specific enzymes, which are biological catalysts. Microorganisms, germinating grain, and man possess various such enzymes.

The chemically reaction groups of sugars arethe hydroxyl groups (-OH) around the ring structure, and upon opening of the ring the

$-\overset{O}{\underset{H}{C}}$ (aldehyde group) and the $-\overset{O}{C}$ (ketone group).

Sugars that possess free aldehyde or ketone groups are known as reducing sugars. All monosaccharides are reducing sugars. Where two or more monosaccharides are linked together through their aldehyde or ketone groups so that these reducing groups are not free, we have nonreducing sugars. The disaccharide maltose is a reducing sugar; the disaccharide sucrose is a nonreducing sugars. Reducing sugars particularly can react with other food constituents, such as the amino acids of proteins, to form compounds that affect the color, flavor, and other properties of foods. In like fashion, the reactive groups of long-chain sugar polymers can combine in a cross-linking fashion. In this case the long chains can align and form fibers, films, and three-dimensional gel-like networks. This is the basis for the production of edible films from starch as a unique coating and packaging material.

Carbohydrates play a major role in biological systems and in foods. They are produced by photosynthesis in green plants. They may serve as structural components as in the case of cellulose; be stored as energy reserves as in the case of starch in plants and liver glycogen in animals; function as essential components of nucleic acids as in the case of ribose, and as components of vitamins such as the ribose of riboflavin. Carbohydrates can be oxidized to furnish energy. Glucose in the blood is a ready source of energy for animals. Fermentation of carbohydrates by yeast and other microorganisms can yield carbon dioxide, alcohol, organic acids, and a host of other compounds.

New Words

derivative 衍生物
glucose 葡萄糖
enzyme 酶
mannose 甘露糖
emulsifier 乳化剂
fermentation 发酵
oxidant 氧化剂
split 使分裂、使分离
pigment 色素
disaccharide 双糖
maltose 麦芽糖
fructose 果糖
amylopectin 支链淀粉
lactose 乳糖
monosaccharide 单糖
polymer 聚合体
dextrin 糊精
polysaccharide 多糖
cellulose 纤维素
amylase 淀粉
hemicellulose 半纤维素
trisaccharide 三糖
pectin 果胶
nucleic acid 核酸

glycogen   糖原
biological catalyst   生物催化剂
gum   树胶
hydroxyl   a. 羟基的
ribose   核糖
aldehyde   醛

Unit 3 Fats

Fats differ from carbohydrates and proteins in that they are not polymers of repeating molecular units. They do not form long molecular chains, and they do not contribute structural strength to plant and animal tissues. Fats are smooth, greasy substances that are insoluble in water.

Fat is mainly a fuel source for the animal or plant in which it is found, or for the animal that eats it. It contains about $2\frac{1}{4}$ times the calories found in an equal dry weight of protein or carbohydrate. Fat always has other substances associated with it in natural foods, such as the fat soluble vitamins A, D, E, and K; the sterols, cholesterol in animal fats and ergosterol in vegetable fats; and certain natural lipid emulsifiers designated phospholipids because of the presence of phosphoric acid in their molecules.

A typical fat molecule consists of glycerol combined with three fatty acids. Glycerol and butyric acid, a common fatty acid found in butter, have the following chemical formulas:

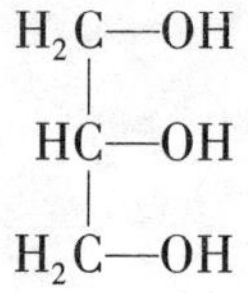

$$\begin{array}{l} H_2C\!-\!OH \\ \ |\ \\ HC\!-\!OH \\ \ |\ \\ H_2C\!-\!OH \end{array} \qquad HOOC\!-\!CH_2\!-\!CH_2\!-\!CH_3$$

Glycerol   Butyric acid

Glycerol has three reactive hydroxyl groups, and fatty acids have one reactive carboxyl group. Therefore three fatty acid molecules can combine with each glycerol molecule, eliminating three molecules of water.

There are about 20 different fatty acids that may be connected to glycerol in natural fats. These fatty acids differ in length and in the number of hydrogen atoms they contain. Formic acid (HCOOH), acetic acid (CH_3COOH), and propionic acid (CH_3CH_2COOH), are the shortest of the fatty acids. Stearic acid ($C_{17}H_{35}COOH$) is one of the longer common fatty acids. Some of the opportunities for variations in natural fats can be seen from the formula for a typical triglyceride:

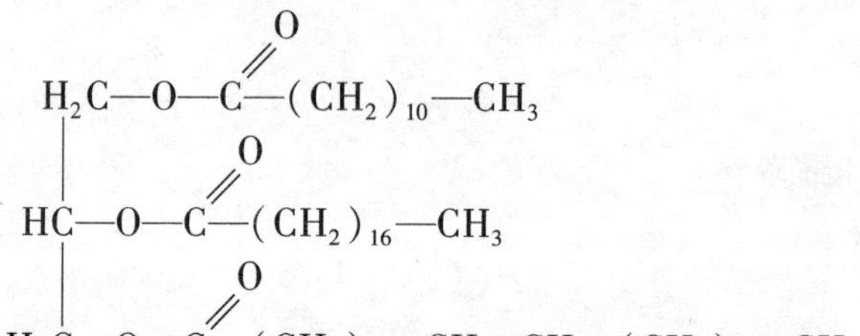

$$\begin{array}{l} H_2C\!-\!O\!-\!\overset{O}{\overset{\|}{C}}\!-\!(CH_2)_{10}\!-\!CH_3 \\ \ |\ \\ HC\!-\!O\!-\!\overset{O}{\overset{\|}{C}}\!-\!(CH_2)_{16}\!-\!CH_3 \\ \ |\ \\ H_2C\!-\!O\!-\!\overset{O}{\overset{\|}{C}}\!-\!(CH_2)_7\!-\!CH\!=\!CH\!-\!(CH_2)_7\!-\!CH_3 \end{array}$$

In this case the fatty acids reacting with glycerol from top to bottom are lauric acid, stearic acid, and oleic acid, with carbon chain lengths of 12, 18, and 18, respectively. Stearic and oleic acids, although of similar length, differ with respect to the number of hydrogen atoms in their chains, stearic acid is said to be saturated with respect to hydrogen. Oleic acid with two fewer hydrogen atoms is said to be unsaturated. Another 18-carbon unsaturated fatty acid with four fewer hydrogen atoms and two points of unsaturation is linoleic acid. This unsaturated fatty acid is a dietary essential for health.

Fat molecules can differ with respect to the lengths of their fatty acids, the degree of unsaturation of their fatty acids, the position of specific fatty acids with respect to the three carbon atoms of glycerol, orientation in the chains of unsaturated fatty acids to produce spatial variations within these chains, and in still other ways.

Fat molecules need not have all three hydroxyl groups of glycerol reacted with fatty acids as in a triglyceride. When two are reacted, the molecule is known as a diglyceride; when glycerol combines with only one fatty acid molecule, the resulting fat is a monoglyceride. Diglycerides and monoglycerides have special emulsifying properties.

Natural fats are not made up of one type of fat molecule but are mixtures of many types, which may vary in any of the ways previously described. This complexity of fat chemistry today is well understood to the point where fats of very special properties are custom-produce and blended for specific food uses.

The chemical variations in fats lead to widely different functional nutritional and keeping quality properties. The melting points of different fats are an example of this functional variation. The longer fatty acids yield harder fats, and the shorter fatty acids contribute to softer fats. Unsaturation of the fatty acids also contributes to softer fats. An oil is simply a fat that is liquid at room temperature, This is the basis of making solid fats from liquid oils. Hydrogen is added to saturate highly unsaturated fatty acids, a process known as hydrogenation.

Some additional properties of fats important in food technology are the following:

They gradually soften on heating that is they do not have a sharp melting point. Since fats can be heated substantially above the boiling point of water, they can brown the surfaces of foods.

When heated further, they first begin to smoke, then they flash, and then they burn. The temperatures at which these occur are known as the smoke point, the flash point, and the fire point. This is important in commercial frying operations.

Fats may become rancid when they are oxidized or when the fatty acids are liberated from glycerol by enzymes.

Fats form emulsions with water and air. Fat globules may be suspended in a large amount of water as in milk or cream or water droplets may be suspended in a large amount of fat as in butter. Air may be trapped as an emulsion in fat as in butter-cream icing or in whipped butter.

Fat is a lubricant in foods——that is, butter makes the swallowing of bread easier.

Fats contribute characteristic flavors to foods and in small amounts produce a feeling of satiety or loss of hunger.

New Words

greasy 油状的
brown 使褐变
flash point 闪点
sterol 固醇
cholesterol 胆固醇
ergosterol 麦角固醇
phospholipid 磷脂
suspend 悬浮
phosphoric acid 磷酸
glycerol 甘油
butyric acid 丁酸
lubricant 润滑剂
formic acid 甲酸
interlace 交织
propionic acid 丙酸
tenderize 使嫩化
stearic acid 硬脂酸
triglyceride 甘油三酯
lauric acid 月桂酸
oleic acid 油酸

Unit 4　Proteins

The molecules of proteins are made up principally of carbon, hydrogen, oxygen, and nitrogen. Most proteins also contain some sulfur and traces of phosphorus and other elements.

Proteins are essential to all life. In animals they help form supporting and protective structures such as cartilage, skin, nails, hair, and muscle. They are major constituents of enzymes, antibodies, and body fluids such as blood, milk, and egg white.

Like carbohydrates, proteins are built up of smaller units called amino acids. These amino acids are polymerized to form long chains. Typical amino acids have the following chemical formulas:

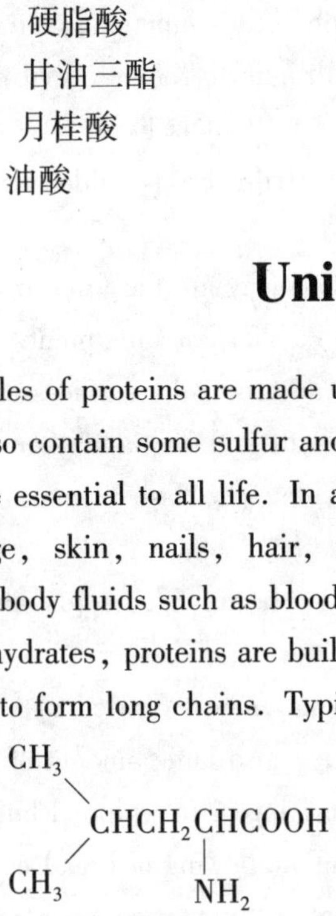

$$\underset{\text{Leucine}}{\overset{\overset{\displaystyle CH_3}{|}}{\underset{\underset{\displaystyle CH_3}{|}}{CH}}CH_2\underset{\underset{\displaystyle NH_2}{|}}{CH}COOH} \qquad \underset{\text{Lysine}}{CH_2\underset{\underset{\displaystyle NH_2}{|}}{CH_2}CH_2CH_2\underset{\underset{\displaystyle NH_2}{|}}{CH}COOH}$$

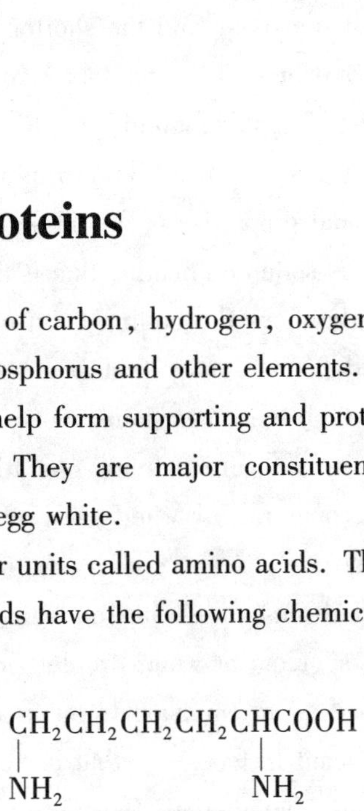

$$\begin{array}{c}CH_3CH_2\\ \diagdown\\ CH_3\diagup^{CHCHCOOH}\\ NH_2\end{array}\qquad\begin{array}{c}CH_3\\ \diagdown\\ CH_3\diagup^{CHCHCOOH}\\ NH_2\end{array}$$

Isoleucine   Valine

Amino acids have the-NH_2 or amino group, and the-COOH or carboxyl group attached to the same carbon atom. These groups are chemically active and can combine with acids, bases, and a wide range of other reagents. The amino and carboxyl groups themselves are basic and acidic, respectively; the amino group of one amino acid readily combines with the carboxyl group of another. The result is the elimination of a molecule of water and formation of a peptide bond, which has the following chemical representation:

$$H_2N-CH-\overset{\overset{O}{\|}}{C}-NH-CH-COOH$$

In this case, where two amino acids have reacted, a dipeptide is formed, with the peptide bond at the center. The remaining free amino and carboxyl groups at the ends can react in like fashion with other amino acids forming polypeptides. These and other reactive groups on the chains of different amino acids can enter into a wide range of reactions with many other food constituents.

There are 20 different amino acids that make up human tissues, blood proteins, hormones, and enzymes. Eight of these are designated essential amino acids since they cannot be synthesized by man at an adequate rate to sustain growth and health and must be supplied by the foods consumed. The remaining amino acids also are necessary for health but can be synthesized by man from other amino acids and nitrogenous compounds and so are designated as nonessential. The essential amino acids are leucine, isoleucine, lysine, methionine, phenylalanine, threonine, tryptophan, and valine. To this list of eight is added histidine to meet the demands of growth during childhood. The nonessential amino acids are alanine, arginine, aspartic acid, cysteine, cystine, glutamic acid, glycine, hydroxyproline, proline, serine, and tyrosine. The list of essential amino acids differs somewhat for other animal species.

There is enormous opportunity for variation among proteins. This variation arises from combinations of different amino acids, from differences in the sequence of amino acids within a chain, and from differences in the shapes the chains assume. That is whether they are straight, coiled, or folded. These differences are largely responsible for the differences in the taste and texture of chicken muscle, beef muscle, and milk curd.

Protein chains can be oriented parallel to one another like the strands of a rope as in wool, hair, and the fibrous tissue of chicken breast. Or they can be randomly tangled like a tangled bunch of string. Thus, proteins taken from different foods such as egg, milk, and meat may have a very similar chemical analysis as to C, H, O, and N, and even with respect to their particular amino acids, yet contribute remarkably different structures to the foods containing them.

Further, the complex and subtle configuration of a protein can be readily changed, not only by chemical agents but by physical means. A given protein in solution can be converted to a gel or precipitate. This happens to egg white when it is coagulated by heat. Or the process can be reversed: a precipitate transformed to a gel or solution as in the case of dissolving animal hoofs with acid or alkali to make glue. This has already been referred to in the case of producing texturized foods from soybean protein.

When the organized molecular or spatial configuration of a protein is disorganized, we say the protein is denatured. This can be done with heat, chemicals, excessive stirring of protein solutions, and acid or alkali.

These changes in food proteins are easily recognized in practice. When meat is heated, the protein chains shrink and so steak shrinks on cooking. When milk is coagulated by acid and heat, protein precipitates, forming cheese curd. If the heat or acid is excessive, the precipitated curd shrinks and becomes tough and rubbery.

Protein solutions can form films and this is why egg white can be whipped. The films hold entrapped air, but if you overwhip you denature the protein, the films break, and the foam collapses.

Like carbohydrate polymers, proteins can be broken down to yield intermediates of various sizes and properties. This can be accomplished with acids, alkalis, and enzymes. The products of protein degradation in order of decreasing size and complexity are protein, proteoses, peptones, polypeptides, peptides, amino acids, NH_3, and elemental nitrogen. In addition, highly odorous compounds, such as mercaptans, skatole, putrescine, and H_2S, may form during decomposition.

Controlled cheese ripening involves a desirable degree of protein breakdown. Putrefaction of meat is the result of excessive protein breakdown accompanying other changes. The deliberate and unavoidable changes in proteins during food processing and handling are among the most interesting aspects of food science. Today animal, vegetable, and microbial proteins are being extracted, modified, and incorporated into numerous manufactured food products. In addition to their nutritional value, they are selected for specific functional attributes including dispersibility, solubility, water sorption, viscosity, cohesion, elasticity emulsifying effects, foamability, foam stability, and fiber formation.

New Words

nitrogen 氮
nitrogenous （含）氮的
sulfur 硫
nonessential 非必需的
phosphorus 磷
phenylalanine 苯丙氨酸
cartilage 软骨
threonine 苏氨酸
nail 指甲
tryptophan 色氨酸

muscle 肌肉
histidine 组氨酸
antibody 抗体
alanine 丙氨酸
body fluid 体液
arginine 精氨酸
egg white 蛋清（白）
aspartic acid 天门冬氨酸
polymerize 使聚合
cysteine 半胱氨酸
isoleucine 异亮氨酸
hydroxyproline 羟脯氨酸
lysine 赖氨酸
proline 脯氨酸
valine 缬氨酸
serine 丝氨酸
carboxyl 羧基的
tyrosine 酪氨酸
base 碱
hemoglobin 血红蛋白
hormone 激素
proteose 朊间质
designate 称为
skatole 粪臭素
mercaptan 硫醇
putrescine 腐胺、丁二胺

Unit 5 Vitamins

Vitamins are chemical compounds in food and needed in very small amounts (in milligrams and micrograms) to regulate the chemical reactions in our bodies and to maintain health. The vitamins are divided into fat-soluble and water-soluble vitamins. Fat-soluble vitamins include vitamins A, D, E, and K. Water-soluble vitamins include the B vitamins and vitamin C. B Vitamins include: thiamin, riboflavin, niacin, vitamin B_6, pantothenic acid, folic acid, biotin, and cobalamin (vitamin B_{12}).

Vitamin A occurs in preformed state and as precursor. Three active forms are retinol, retinal, and retinoic acid. In food, most preformed vitamin A is found in the form of retinal. All three forms of vitamin A can be formed from the plant pigments carotenes. The most common form is beta-carotene. Vitamin A is susceptible to oxidation, but is relatively heat stable. Deficiency of vitamin A results in retardation of growth of young animals. Stoppage of the growth of the skeleton starts early in such deficiency. Skin lesions may occur in man, particularly if the subject is deficient in the B complex vitamins as well. The first symptom in man is night blindness. This is also true for other animals. Although the disease known as xerophthalmia comes later, it is not connected with night blindness. Excessive amounts of vitamin A are toxic. If the doses are too large, nausea, headache, weakness, and dermatitis can result. Green and yellow vegetables are the best sources of the carotenes, the provitamins, while liver is the best for the vitamin. Fish liver oils are the highest and used for medicinal purposes.

The active form of vitamin D is cholecalciferol (vitamin D_3). It can be produced from a protovitamin. It is stored in the liver and functions in the absorption of the minerals calcium and phosphorus. Vitamin D also acts directly on bone, and affects reabsorption of calcium and phosphorus by the kidney. Since rickets is a disease that affects the growing bone, such conditions as knock knee, bowlegs, and similar problems are noticed in children deficient in this vitamin. Excessive administration of the D vitamins can have a toxic effect. Sufficient vitamin D is obtained for children by exposing the body to sunlight from irradiated milk, from vitamin D concentrate of cod liver oil.

Vitamin E or alpha tocopherol is widely available in a normal diet. It functions to detoxify oxidizing radicals that arise in metabolism, to stabilize cell membranes, to regulate oxidation reactions and to protect vitamin A and polyunsaturated fatty acids from oxidation. Sterility occurs in the male rats as a result of tissue damage and failure of movement of spermatozoa, while in the female the lack of this vitamin causes the resorbing of the fetus, Sterility resulting in the male rat is not curable, but in the female the disorder can be cured if the vitamin is included in the food very early in the period of embryonic life. These vitamins are found in seed germ oils and in some vegetables, Nutritional requirements in the United States are supplied by salad oils, shortenings, and margarine. Fruits and vegetables furnish some, as do grain products.

Dietary and intestinal bacterial source contribute to the supply of vitamin K. Storage in the body is minimal, and vitamin K functions in normal blood clotting. If the diet of the chick is deficient in vitamin K, the animal develops a hemorrhagic condition with lengthened clotting time of the blood. Deficiency of vitamin K in the human adult does not occur often because most diets supply sufficient amounts; it is normally produced in the intestine by bacteria as well. A deficiency can occur in newborn infants. This problem results in a hemorrhagic condition which ordinarily disappears when bacteria become established in the intestine. It can be relieved also by administration of vitamin K. Vitamin K is rather widely distributed, and a sufficiency is obtained from the average diet. Excellent sources are alfalfa, cabbage, kale, spinach, and cauliflower. Cereals and fruits are poor sources.

Thiamin functions in carbohydrate metabolism. It makes ribose to form RNA, and it maintains the normal appetite and normal muscle tone in the digestive tract. Thiamin deficiency results in the loss of appetite and development of nausea and later to a form of neuritis and cardiac difficulties. In addition to these, changes in body tissue, blood, and urine take place. Thiamin occurs naturally in yeast, in the bran of rice and wheat, in other cereals, and in the seeds of peas and beans, in peanuts and other nuts, in egg yolk and in pork. Fruits and vegetables usually contain small amounts.

Riboflavin functions as part of a coenzyme involved in oxidation-reduction reactions in energy production. Riboflavin deficiency in man is responsible for glossitis (inflammation) of tongue and lips, and also scaliness at the corners of the mouth (cheilosis). Changes in the eyes which can involve corneal vascularization are also symptoms of riboflavin deficiency. Liver, kidney and heart are good sources of riboflavin, as are milk, bran, wheat germ, egg, cheese, brewers' yeast, and green vegetables.

Niacin functions as a component of two coenzymes involved in oxidation-reduction reactions releasing energy from food. Pellagra is a deficiency disease which causes skin lesions and an inflamed tongue, as well as diarrhea and dementia. Deficiency in the dog results in black tongue. It has been said that this disease causes the three Ds: dermatitis, diarrhea, and dementia——a fourth may be added: dead. Good sources of niacin are yeast, wheat and rice brans, whole cereals, fish, and milk.

Vitamin B_6 was formerly known as pyridoxine. The functions of vitamin B_6 include the metabolism of amino acids and the conversion of glycogen to glucose. Deficiency of pyridoxine has caused convulsion in infants, which can be treated by administration of pyridoxine. Deficiency of these vitamin compounds is not known to be connected with any particular disease in adults. Vitamin B_6 is found in many foodstuffs. Rice bran, yeast, seeds, cereals, egg yolk, as well as meat, liver, and kidney are good sources.

Pantothenic acid is a part of coenzyme A, which is involved in synthesis and breakdown of fats, carbohydrates, and proteins. It is also part of the enzyme, fatty acid synthetase. The recommendation daily allowance is 10 mg a day for adults. A daily intake of 5 ~ 10mg is probably adequate. This vitamin is widely distributed, but the best sources are eggs, liver, and yeast. It is found also in meats and milk.

The coenzyme form offolic acid is tetrahydrofolic acid. It functions in the transfer of formyl and hydroxymethyl group. Folic acid is required for synthesis of purines and pyrimidines and for efficient use of the amino acid histidine. Deficiency of folic acid results in anemia, leukopenia, and failure of growth. Folic acid is widely distributed in green leaves, and liver, yeast, and kidney are known to be good source.

Biotin functions in fatty acid synthesis. Man does not usually suffer from biotin deficiency because bacteria in the intestines are able to synthesize this vitamin in the amounts necessary to maintain health. Minimum daily requirement for biotin has not been established, but a daily intake of 150 ~ 300μg is considered adequate. Biotin is widely distributed in nature. Peanuts, egg yolks, beef liver, chocolate, and yeast are good sources of biotin.

Cobalamin or vitamin B_{12} is required for nucleic acid synthesis, amino acid synthesis, blood cell formation, neural function, and growth. A patient suffering from pernicious anemia can be successfully treated by intramuscular injection of this vitamin. Cobalamin is found only in animal products. Liver and kidney are excellent sources of vitamin B_{12} and it is present also in meat, milk, fish, and poultry.

Ascorbic acid or vitamin C functions in wound healing, collagen synthesis, iron absorption, and as an antioxidant. Vitamin C is necessary for conversion of proline to hydroxyproline and lysine to hydroxylysine. It is involved in iron absorption and the conversion of amino acids to neurotransmitter. It is the least stable of all vitamins. It oxidizes readily in light or air, when heated or in alkaline solution. Degradation is enhanced by presence of iron and copper. Scurvy is a disease resulting from the lack of ascorbic acid in the diet. Man, monkeys, the guinea pig, the Indian fruit bat, and fish are unable to synthesize this vitamin and must obtain it from dietary sources. Ascorbic acid is widely distributed in the plant kingdom, being especially prevalent in citrus fruits, black currants, strawberries, pineapples, broccoli, and parsley.

New Words

riboflavin 核黄素,维生素 B_2
pantothenic acid 泛酸
folic acid 叶酸
biotin 生物素,维生素 H
cobalamin 钴胺素,维生素 B_{12}
retinol 视黄醇,维生素 A
retinal 视黄醛,维生素 A 醛
retinoic acid 视黄酸,维生素 A 酸
xerophthalmia 眼干燥,干眼病
dermatitis 皮炎
provitamins 维生素原,前维生素
cholecalciferol 胆钙化【甾】醇,维生素 D_3
protovitamin 维生素原
rickets 佝偻病,软骨病
knock-knee 膝外翻
bowleg 向外弯曲之腿,O 型腿,
tocopherol 生育酚
detoxify 除去……之毒物,使解毒
spermatozoon 精虫,精子
resorbing of the fetus 吸收胎
embryonic 胚胎的,幼虫的
hemorrhagic 出血的
alfalfa 紫花苜蓿
kale 无头甘蓝
ribose 核糖
tone 身体(各器官)的健康状态
neuritis 神经炎

cardiac 心脏的,心脏病的
coenzyme 辅酶
glossitis 舌炎
scaliness 鳞片状,鳞片
cheilosis 唇损害,唇干裂
corneal 角膜的
vascularization 血管化,微管化
brewers, yeast 啤酒酵母
pellagra 糙皮病
diarrhea 腹泻
dementia. 痴呆,智力衰失
glycogen 糖原,肝糖
convulsion 痉挛
synthetase 合成酶
allowance 容许量,许可限度
tetrahydrofolic acid 四氢叶酸
formyl 甲酰基
hydroxymethyl 羟甲基
purines 嘌呤
pyrimidines 嘧啶
leukopenia 白血球减少症
neural 神经的
pernicious anemia 恶性贫血
intramuscular 位于肌肉内的
collagen 胶原
antioxidant 抗氧化剂
neurotransmitter 神经传递素
scurvy 坏血病
guinea pig 豚鼠,天竺鼠
fruit bat 果蝠,狐蝠,犬蝠

Unit 6　Minerals

　　Mineral substances play an important part in the nutritional value of foods. They are usually present in small amounts, frequently in the traces, but their importance is well recognized. Minerals are present in the form of salts of metals or in combination with organic compounds such as phosphoproteins and enzymes containing metals.

　　The major elements include potassium, sodium, calcium, magnesium, chlorine, sulfur, and phosphorus. Trace elements include iron, copper, iodine, cobalt, fluorine, and zinc. The role of selenium in the human body is not completely established. Others of unknown nutritive value include aluminum, boron, chromium, nickel, and tin. Toxic elements include arsenic, cadmium, mercury, lead, and antimony. Food materials are the important source of the major and minor elements.

Occurrence of Minerals
Minerals in Major Amounts

Calcium: An important source of calcium in the diet is milk. Cereals and other foods make a contribution of calcium to the diet also. Calcium deficiency is noted in diets low in protein and other major nutrients. The period of its greatest needed is during pregnancy, lactation, and growth. Tests with animals have shown that larger amounts of calcium are retained when lactose is added to the diet.

Sodium phytate (hexaphosphoinositol) tends to from insoluble salts with calcium in theintestines. This makes calcium unavailable for absorption. This compound is present in cereals and soybeans. The problem can be overcome by a sufficiency of calcium intake.

Calcium salt in small amounts is frequently added to tomatoes during the canning process to increase the firmness. The compound formed is calcium pectate (Loconti and Kertesz 1914).

Magnesium: All plants and animals require magnesium. Chlorophyll supplies the largest amount of magnesium in the adult diet. Infants are given supplementary amounts of magnesium because milk does not have an adequate supply.

Magnesium is important to cardiovascular function, including involvement in blood pressure, myocardial contraction, and myocardial conduction and rhythm. It is important in oxidative phosphorylation, enolase, and glucokinase.

Sodium: Sodium chloride as well as sodium present in foods of animal and vegetable origin is the source of this element in the diet. Water supplies contain considerable amounts of sodium but not as the chloride.

Salt is important in the diet. However, it is thought by some that the average intake of salt in theUnited States is too high. Sodium is important in the preservation of pH the body fluid and in body fluid volume.

Potassium: Potassium is present in plant materials and ingestion of these plant materials is a source of potassium in the diet. Potassium is important in the maintenance of the pH of body fluids, as well as in water and electrolyte balance in these fluids. K^+ has a role in enzyme systems.

The infant protein malnutrition condition known as kwashiorkor has been found to affect the K^+ levels. This involves as deficiency of dietary amino acids. The result is disturbance of electrolyte and water balances and a loss of potassium from the cells. Such a loss is made worse by the diarrhea which accompanies it.

Under experimental conditions rats red diets deficient in potassium showed slow growth with renal hypertrophy and with other symptoms. The final result was death.

Minerals in Minor Amounts

Trace elements are found everywhere and are present in all natural foods. Under ordinary circumstances the quantity of a trace element in a given food is related to the amount of this element found in the environment. It is obvious that these elements arise from the soil in the case of plants and from feed in the case of animals. These elements can also get into foods from the processing equipment. Some animals such as sword fish have the capability to absorb mercury from the seawater. Such absorption has made the flesh of this species undesirable for food when the mercury level is too high.

Iron: Foods which supply large amounts of iron include liver, kidney, and heart, as well as egg yolk, shellfish, cocoa, parsley, and molasses. Muscle, green vegetable, nuts, poultry, whole wheat flour, and bread supply smaller amounts.

Iron is ingested ordinarily in the form of iron porphyrin or iron-protein complexes. The use of iron salts in the diet has long been practiced for the beneficial effect. The hemoglobin of red cells, cytochromes of respiration, and myoglobin of heart muscle require iron for normal operation. Iron is also involved in activity of the enzymes catalase and peroxidase. The complexing of iron by the porphyrin nucleus concerns the bulk of these situations. Increased assimilation of iron is helped by diets high in ascorbic acid and low in phosphates.

Copper: It has been only during recent years that copper has been found to be involved in human normal and abnormal physiology. Other metals and anions influence fluid retention and metabolism. In addition, copper is a component of a number of enzymes. A dietary balance of copper and other minerals is necessary.

Good dietary sources of copper include liver, kidney, shellfish, including oysters and crustaceans, nuts, cocoa, peaches, and grapes. Fresh fruits, non-leafy vegetables, and refined cereals contain small amounts of copper.

Cobalt: Vitamin B_{12} contains cobalt. This is discussed in unit 5 Vitamins. This vitamin is primarily connected with pernicious anemia. A patient suffering from this disease fails to absorb the vitamin from the intestinal tract. It is not just dietary inadequacy. The cause is the lack of an intrinsic factor in the gastric juice. This factor is a mucoprotein which binds with a molecule of vitamin B_{12}. It can then be used by the body. Vitamin B_{12} is found in protein foods of animal origin of which liver and kidney are the best sources.

Manganese: The role of manganese in essential enzyme reactions is well known. Enzymes which require Mg^{2+} or Mn^{2+} can usually act with Mn^{2+}. Pyruvate kinase functions with either Mg^{2+} or Mn^{2+}. It forms a complex with either of these metals before it binds with the substrate. A deficiency of this metal has not been known in humans. Liver, heart, and kidney contain relatively high amounts of manganese. Nuts and cereals are highest. Meranger and Somers (1968) give much information on the manganese values of seafood. Other metals are included in this report also.

Zinc: A fair amount of zinc is present in the human body. It is involved in the healing of wounds, and there seems to be a connection between the levels of zinc and the production and action of insulin. It seems likely that zinc is essential for a large number of physiological processes.

It is present in some enzymes. Alcohol dehydrogenase of yeast and glutamic acid dehydrogenase of beef liver contain zinc in the protein molecule.

Bran and wheat germ, as well as oysters, are high in zinc. The former contain $40 \sim 120 \mu g/g$, while oysters contain 1000μg/g. White bread, meat, fish, leafy vegetables, nuts, eggs, and whole cereals contain moderate amounts of zinc.

Acid foods should not be stored in galvanized containers, because it is possible to dissolve enough zinc to cause poisoning.

Molybdenum: Molybdenum seems to have an effect on dental caries. It was found that molybdenum plus fluoride is more effective than fluoride alone.

The metabolic enzyme xanthine oxidase contains molybdenum in the molecule. It catalyzes the oxidation of xanthine to uric acid.

Molybdenum is found in large amounts in liver kidney, legumes, leafy vegetables, and cereal grains. Cereal grains contain from 0.12 to 1.114μg of molybdenum while dried legume seeds contain 0.2~4.7μg/g. Fruits, meats, foot and stem vegetables, and milk are poorest in this metal.

New Words

mineral substances　矿物质
organic compounds　有机化合物
phosphoproteins　磷蛋白质【生化】磷朊
potassium　【化学】钾
sodium　【化学】钠
calcium　【化学】钙
magnesium　【化学】镁
chlorine　【化学】氯
sulfur　硫磺；硫磺色；用硫磺处理
phosphorus　磷
trace elements　微量元素
iron　铁
copper　铜
iodine　碘
cobalt　【化学】钴
fluorine　【化学】氟
zinc　锌
selenium　【化学】硒
nutritive value　营养价值
aluminum　铝
boron　硼
chromium　【化学】铬
nickel　镍
tin　锡
toxic elements　有毒元素

arsenic 砷
cadmium 镉
mercury 水银;水银柱
lead 铅
antimony 【化学】锑
minor elements 微量元素
nutritional function 营养功能
physiologically 生理学方面
Co cobalt 【化学】钴
Cu copper 铜
Zn zinc 锌
lactose 乳糖
sodium phytate 【有机化学】植酸钠
insoluble salts 不溶性盐
intestines 内脏
cereals 谷类;谷类食品
soybeans 大豆;黄豆
calcium salt 钙盐
pectate 【生化】果胶酸盐
chlorophyll 叶绿素
adult diet 成人饮食
cardiovascular 【解剖】心血管的
blood pressure 血压
myocardial contraction 心脏收缩
conduction 【生理】传导
oxidative phosphorylation 【生化】氧化磷酸化
enolase 【生化】烯醇酶
glucokinase 【生化】葡糖激酶
sodium chloride 【无化】氯化钠
chloride 氯化物
average intake 平均摄入
body fluid 【生理】体液
electrolyte balance 【医学】电解质平衡
enzyme systems 【生化】酶系统
malnutrition 营养失调;营养不良
electrolyte 电解液;电解质
water balances 水分平衡;液体平衡
experimental conditions 实验条件
renal 【解剖】肾脏的

hypertrophy 【病理】肥大;过度增大
trace elements 微量元素
natural foods 天然食品
processing equipment 加工设备
sword fish 刀鱼
kidney 【解剖】肾脏;腰子
egg yolk 蛋黄
shellfish 甲壳类动物;
cocoa 可可粉;可可豆
parsley 荷兰芹;欧芹
molasses 糖蜜;糖浆
muscle 肌肉
poultry 家禽
whole wheat flour 全麦粉
porphyrin 【生化】卟啉
iron-protein complexes 铁蛋白质复合体;铁蛋白质络合物
hemoglobin 【生化】血清蛋白
cytochromes 细胞色素
respiration 呼吸作用
myoglobin 肌红蛋白
heart muscle 心肌
normal operation 正常运作
catalase 过氧化氢酶
peroxidase 过氧化物酶
complexing 络合;配位
porphyrin 【生化】卟啉
assimilation 【生化】同化作用
ascorbic acid 抗坏血酸
phosphates 磷酸盐
physiology 生理学;生理学机能
oysters 牡蛎,蚝(oyster 的复数形式)
crustaceans 甲壳类;甲壳纲动物
nuts 坚果
vitamin 维他命;维生素
pernicious anemia 【内科】恶性贫血
intrinsic factor 内在因素
gastric juice 胃液
mucoprotein 粘蛋白
animal origin 动物来源

enzyme reactions　酶反应
substrate　基质
healing　康复
wounds　创伤
insulin　胰岛素
physiological processes　生理过程
alcohol dehydrogenase　乙醇脱氢酶
yeast　酵母
glutamic acid dehydrogenase　谷氨基酸脱氢酶
wheat germ　小麦胚芽；麦芽
galvanized　镀锌的，电镀的（galvanize 的过去式和过去分词形式）
dissolve　溶解；使溶解
poisoning　中毒
molybdenum　钼
dental caries　蛀牙
fluoride　氟化物
metabolic　变化的；新陈代谢的
xanthine oxidase　黄嘌呤氧化酶
catalyze　催化
oxidation　氧化
xanthine　黄嘌呤
uric acid　尿酸
cereal grains　粮谷
legume　豆类
essential nutrients　基本营养物；必须营养素
fatty acids　脂肪酸
fiber　纤维
specific function　特殊功能
digestive system　消化系统

Unit 7　Enzymes

Enzymes are catalysts produced by living cells that regulate the speed and specificity of the thousands of chemical reactions that occur in protoplasm. An enzyme does not have to be inside a cell to act as a catalyst: many of them have been extracted from cells with their activity unimpaired. They can then be purified, crystallized and their catalytic abilities can be studied. Enzyme-controlled reactions are basic to all the phenomena of life: respiration, photosynthesis, nitrogen fixation, deamination, digestion, and so on.

All the enzymes isolated and crystallized to date have been found to be proteins. They are usually colorless, but they may be yellow, green, blue, brown or red. Most enzymes are soluble in water or dilute salt solution, but some, for example, the enzymes present in the mitochondria, are bound together by lipoprotein (a phospholipid-protein complex) and are insoluble in water. Enzymes are usually named by adding the suffix—ase to the name of the substance acted upon, e.g., sucrose is acted upon by the enzyme sucrase.

The earliest clue and ideas about enzymes were connected with work on digestion and fermentation. In the seventeenth and eighteenth centuries experiments and observations suggested that these two processes were somewhat similar in nature. Early in the 19th century it was shown that starch is converted to glucose in the presence of dilute acid, during which process the acid involved is not altered, and can be recovered unchanged. It was found, also, that something in gastric juice degrades proteins. These and other experiments brought about the idea of catalysis in which the substances acting as agents for the change are in themselves unchanged. It was still thought that life itself in some mysterious way brought about fermentation. Late in the nineteenth century, Buchner, after breaking up yeast cells, showed that a solution made from yeast from which all the cells were removed by special filtration under pressure could cause fermentation of sugar to CO_2 and alcohol. This showed that living cells were not necessary to produce fermentation. The connection between enzymes and fermentation explains why enzymes were formerly called ferments.

Sumner first isolated an enzyme, urease, in the crystalline form for which he won a Nobel prize. After Sumner's achievement, Northrop prepared other enzymes in crystalline form and extended Sumner's concept that enzymes are proteins. Northrop also was awarded the Nobel prize.

The catalytic ability of some enzymes is truly phenomenal. For example, one molecule of the enzymes catalase, extracted from beef liver, will bring about the decomposition of 5,000,000 molecules of hydrogen peroxide (H_2O_2) per minute at 0℃. The substance acted upon by an enzyme is known as its substrate; thus, hydrogen peroxide is the substrate of the enzyme catalase. The number of molecules of substrate acted upon by enzyme per minute is called the turnover number of enzyme. The turnover number of catalase is thus 5,000,000. Hydrogen peroxide is a poisonous substance and is produced as a by-product in a number of enzyme reactions. Catalase protects the cell by destroying the peroxide. Most enzymes have high turnover number, which explains why they can be so effective although present in such minute amounts. It is this catalytic efficiency of enzymes at low temperature that makes them important to the food scientist. This means that foods can be processed or modified by enzymes at moderate temperature, say 25~50℃, where food products would not otherwise undergo changes at a significant rate. It also means, however, that endogenous enzymes are active under these conditions as well, and this can be beneficial or deleterious.

Of course, one basis for heat processing is to denature and inactivate enzymes so that the food is not subjected to continuing enzymic activity. The food scientist must have an understanding of the denaturation phenomenon in order to properly process foods.

Enzymes differ in their specificity, in the number of different substrates they will attack. A few enzymes are absolutely specific: urease, which decomposes urea to ammonia and carbon dioxide, will attack no other substance, and a specific enzyme is necessary to split each of the three common double sugars sucrose, maltose and lactose. Amylase, adigestive enzyme, will hydrolyze starch, but not cellulose. Both molecules are polymers of glucose. They differ in the orientation of one bond at the junction of glucose. Other enzymes are relatively specific and will work upon only a few, closely related substances. Peroxidase will decompose several different peroxides including hydrogen peroxide. Peroxidase is found in a wide variety of plant and animal tissues.

Finally, a few enzymes are specific only in requiring that the substrate have a certain kind of chemical bond. The lipase secreted by the pancreas will split the ester bonds connecting the glycerol and fatty acids of a wide variety of different fats. It is this enzymic specificity that allows the food scientist to selectively modify individual food components and no affect others.

The sensitivity and specificity of enzymes also make them important to the food scientist as analytic tools. Analysis for food components in many instances can be simplified using enzymic techniques.

Some enzymes, such as pepsin, consist solely of protein. Many others consist of two parts, one of which is protein (called an apoenzyme) and the second (called a coenzyme) is made of a smaller organic molecule, usually containing phosphate. Coenzymes can usually beseparated from their enzymes and, when analyzed, have proved to contain some vitamin as part of the molecule—thiamine, niacin, riboflavin, pyridoxine, and so on. This finding has led to the generalization that all vitamins function as parts of coenzymes in body cells. Neither the apoenzyme nor the coenzyme alone has any catalytic activity; only when the two are combined is activity present. Other enzymes require for activity, in addition to a coenzyme, the presence of some ion. Several of the enzymes involved in the breakdown of glucose require magnesium. Ptyalin, the enzyme of saliva, requires chloride ion for activity. Most, if not all, of the elements needed by plants and animals in very small amounts—the so-called trace elements, manganese, copper, cobalt, zinc, iron, etc. —function as such enzyme activators.

Enzymes are classified according to the reactions they catalyze. In some cases, the terms used are fairly clear; in others, less so. Examples are oxidoreductases, transferases, hydrolases, isomerases, lyases, and ligases.

Enzymes are important in a number of industrial processes. Enzymes can be advantageous because of the following characteristics. (a) They areefficient catalysts. (b) The reaction can be stopped by applying heat sufficient to destroy the enzymes. (c) Temperature, pH, and time can be used to control the reactions. (d) They are nontoxic and can be left in the product unless it is necessary to destroy them.

The activity of enzymes is of considerable importance in food technology, since many such reactions can have beneficial results and others cause undesirable changes. It is necessary that these facts be taken into consideration in the storage and processing of foods. Enzymes can be important in the production of flavors in fruits and vegetables. Enzymes are also of importance in the production of tender meat.

New Words

enzymes 酶
protoplasm 原生质,细胞质
nitrogen fixation 固氮作用
deamination 脱氨作用
mitochondria 线粒体
lipoprotein 脂蛋白
phospholipid 磷脂
sucrose 蔗糖
sucrase 蔗糖酶
glucose 葡萄糖
protein 蛋白质
urease 尿素酶
phenomenal 显著的,不寻常的,现象的
catalase 过氧化氢酶
hydrogen peroxide 过氧化氢
substrate 底物
turnover number 转换数
denature 变性
inactivate 钝化
specific 专一的
urea 尿素
ammonia 氨
carbon dioxide 二氧化碳
maltose 麦芽糖
lactose 乳糖
amylase 淀粉酶
cellulose 纤维素
peroxidase 过氧化物酶
chemical bond 化学键
pancreas 胰腺
ester bonds 酯键
glycerol 甘油
fatty acids 脂肪酸
apoenzyme 酶蛋白,辅酶蛋白质部分
coenzyme 辅酶
thiamine 硫胺素,维生素 B_1

niacin　烟酸,尼克酸,维生素 PP
pyridoxine　吡哆醇,维生素 B_6
magnesium　镁
ptyalin　唾液淀粉酶
saliva　唾液
chloride　氯
trace　痕量
manganese　锰
cobalt　钴
activator　激活剂
oxidoreductase　氧化还原酶
transferase　转移酶
hydrolase　水解酶
isomerase　异构酶
lyases　裂合酶,裂解酶
ligase　接合酶,连接酶

Unit 8　Nutrition

People require energy and certain other essential nutrients. These nutrients are essential because the body cannot make them and must obtain them from food. Essential nutrients include vitamins, minerals, certain amino acids, and certain fatty acids. Foods also contain other components such as fiber that are important for health. Although each of these food components has a specific function in the body, all of them together are required for overall health. The digestive system breaks down food into nutrients for absorption.

Nutrient Needs

In the United States the nutrition needs of the public are estimated and expressed in the Recommended Dietary Allowances (RDA). These were initially established during World War Ⅱ to determine in a time of possible shortage, what levels of nutrients were required to ensure that the nutrition of the people would be safeguarded. The RDA are established by the Food and Nutrition Board of the National Research Council, whose members come from the National Academy of Sciences, the National Academy of Engineering, and the Institute of Medicine.

The first RDA were published in 1943 by a group known as the National Nutrition Program, a forerunner of the Food and Nutrition Board. Initially, the RDA were intended as a guide for planning and procuring food supplies for national defense. Now RDA are considered to be goals for the average daily amounts of nutrients that population groups should consume over a period of time.

The RDA are the levels of intake of essential nutrients considered, in the judgment of the Food and Nutrition Board on the basis of available scientific knowledge, to meet the known nutrition needs of practically all healthy persons. The NAS-NRC recognizes that diets are more than combinations of nutrients and should satisfy social and psychological needs as well.

As the needs for nutrients have been clearly defined, the RDA have been revised at roughly five-year intervals. The Ninth Edition of the RDA was published in 1980. The Tenth Edition was due to be released in 1986, but controversy regarding some of its recommendations delayed its publication until 1989.

The requirement for a nutrient is the minimum intake that will maintain normal functions and health. In practice, estimates of nutrient requirements are determined by a number of techniques, including:

Collection of data on nutrient intake from apparently normal, healthy people.

Determinations of the amount of nutrient required to prevent disease states (generally epidemiological data).

Biochemical assessments of tissue saturation or adequacy of molecular function.

A major revision is in process to replace the RDA. The revised recommendations are called Dietary Reference Intake (DRI). Until 1997, the RDA were the only standards available. They will continue to be useful until DRI can be established for all nutrients.

Food Pyramid

Recently, the FDA has proposed the Food Pyramid to help consumers determine what they should eat to meet their dietary needs and to prevent disease. The following is suggested:

—Bread, cereal, rice, and pasta-6 to 11 servings

—Vegetable-3 to 5 servings

—Fruits-2 to 4 servings

—Milk, yogurt, and cheese-2 to 3 servings

—Meat, poultry, fish, nuts, and beans-2 to 3 servings

—Fats, oils, and sweeteners-use sparingly

The USDA also has made some recommendations and provides a set of guidelines for healthy eating. These last two are generally much easier for the average person to follow and ensure adequate nutrition in the diet.

Bioavailability of Nutrients

Chemical analysis of a food may determine the presence of a nutrient, but this can be misleading. Though the nutrient may be in the food, whether it is available in a form that can be used by the metabolic processes of the body is another question. If the nutrient is in a form that can be used, it said to be bioavailable. Factors determining the bioavailability of a nutrient include digestibility, absorption, nutrient-to-nutrient interactions, binding to other substances, processing and cooking procedures. Also, age, gender, health, nutritional status, drugs, and food combinations influence the bioavailability of carbohydrates, proteins, fats, vitamins, and minerals.

Stability of Nutrients

The nutritive value of food starts with the genetics of the plants or animals. Fertilization, weather, and the maturity at harvest also influence the composition of the plant or animal being used for food. Storage before processing affects nutrient levels. Then all of the processing steps continue to affect the nutrient levels in a food. Finally, preparation in the home or at the restaurant can reduce the final nutritive value of a food before the digestive process.

A primary goal of food science is to preserve the nutrients through all phases of food harvesting, processing, storage, and preparation. To do this, the food scientist needs to know what the stability of the nutrients is under varying conditions of pH, air, light, heat, and cold. Nutrient losses are small in most modern food processing operations, but when nutrient losses are unavoidably high, the law allows enrichment.

New Words

RDA 推荐日摄食量
National Research Council 国家研究委员会
National Academy of Sciences （美国）国家科学院
national defense 国防
minimum intake 最小摄入量
normal functions 正函数
nutrient requirements 营养需要量
generally epidemiological data 一般流行病学数据
biochemical assessments 生物学评价
dietary reference intake 膳食参照摄入量
food pyramid 食物金字塔
pasta 意大利面食
yogurt 酸奶酪
poultry 家禽
sweeteners 甜味剂
blood clotting 血液凝集
gland 腺
hemoglobin 血红蛋白；血红素
skin pigments 皮肤色素
collagen 胶原；胶原质
enzyme activator 酶催化剂
energy metabolism 能量代谢
tissue fluid 组织液
acid-base balance 酸碱平衡
USDA 美国农业部
bioavailability 生物利用度；生物药效率
metabolic processes 代谢过程
digestibility 消化性；可消化性
nutrient-to-nutrient interactions 养分的相互作用

nutritional status   营养状况
carbohydrates   糖类;碳水化合物
nutritive value   营养价值
genetics   遗传学
fertilization   【生】受精;施肥;肥沃
processing steps   加工步骤;处理步骤
food processing operations   食品加工操作

Exercises

1. How many ingredients in foods?
2. What are the roles of water? What is the definition of A_w?
3. How many types of carbohydrates? Please write out their names.
4. What are the roles of carbohydrates?
5. What is the definition of protein? Which elements are composed of protein?
6. What is the structure of Amino acids?
7. How many amino acids are made up human tissues, blood proteins, hormones, and enzymes?
8. What are the roles of fats? What is the structure of fats?
9. What are vitamins? What are the roles of them?
10. What is enzyme? Please write out their structures and roles.
11. Name six minerals required by the body.
12. Describe the function of protein in the diet.
13. What nutritional deficiency caused kwashiorkor and marasmus?
14. What factor determines protein quality?

Chapter 3　Food Technology

Unit 1　Sterilized Milk and Milk Products Cultured Milk Products

Fermented milk products including cheeses, yogurt, buttermilk, and sour cream have formed a vital segment of the human diet in many regions of the world. Modern production processes use defined cultures for fermenting milk of the cow, goat, and other dairy animals to yield nutritious and wholesome cultured dairy products. Present day processes give predictable results in contrast to spontaneous fermentation by inherent or environmental microflora in earlier times. Modern fermentation technological procedures can generally be directed because they are carried out under specific conditions of temperature, pH, composition of milk base, and discrete microbial culture. Accordingly, the end products are practically free of public health hazards and generally conform to the attributes and quality associated by the consumer with the particular product.

The fermentation processes extend the shelf-life of milk nutrients, giving thereby a marketing and distribution capability to cultured dairy products. Nutritional quality of cultured milk products is considerably enhanced since the cultures synthesize B-vitamins and modify milk proteins, lactose, and lipids. In addition, the cultures have been said to produce certain metabolites such as antibiotics, anticarcinogenic compounds, and an anticholesteremic factor, which render cultured milks evidently beneficial from a therapeutic standpoint. Furthermore, the microbial lactases of lactic cultures in cultured dairy foods aid in lactose digestion. Consequently, lactose-intolerant individuals can digest milk significantly better in the presence of cultures in cultured milk foods. Recently it has been documented that rats fed yogurt gained weight faster than those fed unfermented milk or acidified milk.

As a result of the growth of the lactic cultures, different textures and flavors are generated in milk. Major factors governing the flavor and texture development are composition of the dairy mix and nature of the culture employed. All the cultured foods except buttermilk have shown growth in recent years. The market has boomed for yogurt since the introduction of fruit-flavored yogurts in the mid-1960. It appears that yogurt is enjoying unprecedented growth because of its image as a convenient, health oriented, and low fat food. More than 500 million lb of yogurt are presently consumed and the trend is towards increasing consumption in the foreseeable future. The buttermilk market is declining, although in volume this product is twice as big as yogurt. The market for sour cream is rising much faster than that for cottage cheese. Cheeses (other than cottage cheese) have also enjoyed a significant growth in recent years.

The lactic cultures are nonspore formers. They may be purchased from a commercial culture manufacturing organization. The starters are purchased for a specific product. Generally, the starters contain several strains of a single species of lactic bacteria. Since bacteriophage attacks a specific strain, only a mixture of strains insures dependable fermentation capability of the starter. Bacteriophage is also combated by scrupulous plant sanitation and the use of phosphated media for bulk starter preparation.

Direct acidification procedures (avoiding the use of bacterial cultures) for buttermilk, cottage cheese, and sour cream have been recently developed. The B-vitamin content of directly acidified products is generally lower than their fermented counterparts.

In addition, the major by-product of the cheese industry, whey and its fractions, is being investigated as a means of producing economically viable products for application in feed and food industries. Whey fermented with *L. bulgaricus* and subsequent neutralization with ammonia and dehydration yields an effective nitrogen supplement in the nutrition of dairy animals. Single cell protein concentrates have been from whey fermented with lactose-hydrolyzing yeasts, which can be used prepared as well as for animals. In addition, whey can be used effectively for the manufacture of industrial alcohol for blending into gasohol. A process has been developed which involves recovering proteins from whey by ultrafiltration and fermenting the lactose in whey permeate into alcohol by a single or combined culture of *Saccharomyces cerevisiae* and *Kluyveromyces fragilis*. Alternatively, the whey permeate can be mixed with corn and the combined sugars in the whey: grain mixture fermented into alcohol.

New Words

culture　培养体,发酵剂
gasohol　气孔
vital　必不可少的,极其重要的
ultrafiltration　超速过滤
spontaneous　出自自然的、不依赖人工的
saccharomyces cerevisiae　啤酒酵母
inherent　体内的、自身的
Kluyveromyces fragilis　克罗氏脆壁酵母
modify　溶解麦芽糖
anticarcinogenic　抗致癌物的
anticholesteremic　抑制胆固醇形成的
therapeutic　治疗学、治疗
acidify　酸化、用酸调
image　象征、反映
bacteriophage　噬菌体
acidophilin　嗜酸链乳杆菌素
Lactobacillus acidophilus　嗜酸乳杆菌
L. bulgarlcus　保加利亚杆菌

Unit 2　Cheese

Cheese is a selectively concentrated, highly nutritious food which can be made from the milk of most mammals. The milk of four are important—cow, sheep, goat, and buffalo—with the first dominating world cheese production. Many countries are heavy producers of cheese.

The beginnings of cheesemaking occurred perhaps about 5000 B. C. in the Mesopotamia area of Southwest Asia, but its technology evolved slowly until the second half of the 20th century, when per capita consumption increased to high levels. Flavor, consistency, nutritional value, and easy availability were contributing factors. Among the basic cheeses of the world, Cheddar is the most utilized in the U. S. followed by Cottage cheese.

Two major classifications of cheese exist—fresh, unripened as Ricotta, Cream, and Cottage, and ripened as Swiss, Cheddar, Camembert, Blue mold, and Brick—each showing different compositions.

Both classes utilize bacterial cultures to initiate the necessary lactic acid fermentation, but milk for fresh, unripened cheeses is precipitated or curdled either by developing enough lactic acid to attain the isoelectric point of its casein, like Cottage, or by attaining a critical combination of lactic acid development and high heat which precipitates the casein at a pH higher than its isoelectric point, like Ricotta.

In the ripened cheese classification, casein of the cheesemilk is transferred to an unstable state through the complex activity of an added enzyme preparation containing proteases such as rennin, which hydrolyze the k-casein. This leads to the rapid formation of a smooth sweet curd (pH 6.2) in pathways postulated by various investigators.

About eight major steps are involved in cheesemaking for either fresh or ripened cheese, but different type cheeses emerge because of the degree of emphasis placed on each step, the differences in environment, and, to some extent, the type of microorganism introduced.

After the curd is formed, cut, and cooked, the whey is separated and the curds are salted. Special steps then follow, in most cases involving the application of unique microorganisms or the controlling of the environment, which with time induces unique changes in flavor, texture, and appearance.

Ripening refers to a period and a place where the prepared salted curds, pressed into specific forms or shapes, are purposely held at a predetermined temperature, ranging from 5℃ to 15℃, and relative humidity of 85% to 95%, until the traits characteristic for a certain type of cheese are attained. At this point the mature cheeses are removed and cleaned, and may be cut and packaged into smaller consumer sizes. They may be stored at 4℃ or lower or distributed immediately. Process cheeses can be made from these natural ripened cheeses by grinding them, introducing fat and protein emulsifying salts, and heating to about 70℃.

Among the most interesting natural cheeses are the mold-derived cheeses such as Blue, Roquefort, Camembert, and Brie. The first two require the special applications to milk or curds of the blue-green mold spores of *Penicillium roquefocti* or *P. glaucum*. These grow and produce mycelia or strands throughout the cheese, as soon as air channels are created by puncturing the wheel with metal needles. Roquefort is produced in only one region of southern France, and only from sheep milk. The second two cheeses require the special application to the cheesemilk, or to the surfaces of fresh wheels, or both, of the mold spores of *P. caseiocolum* or *P. camemberti*. Here on the surfaces of the cheeses only a white mycelia mat develops, as this mold grows well in open air. Mold mycelia, whether for blue or white cheeses, introduces active lipolytic and proteolytic enzymes to the cheese, which, in consort with existing enzyme systems in milk, and with those of the natural yeast growing in a parallel time sequence on the mold cheese surface, give the cheese its characteristic flavor and texture at about 10℃ and 95%, relative humidity.

Cheesemaking has been an art, but now Czulak and others show mechanization of cheesemaking is making great strides. But still in the Alpine areas, and elsewhere, traditional methods leading to outstanding cheeses prevail, even to utilizing directly the freshly macerated rennets of calf stomachs, rather than commercially prepared liquid or powdered rennet.

Continuous cheescmaking has been advanced by the new ultrafiltration concept of Maubois and Mocquot, where milk is selectively concentrated to a precheese state, at which point lactic acid culture, rennet, color, salt, and mold spores are metered into the plastic mass. The resulting curd without significant free whey forms quickly as fresh cheese, which is then conventionally ripened. New cheesemaking methods arising from this concept have been reported. Advantages include significantly increased cheese yields, lower requirements for rennet, and more continuity in operations. In France much fresh unripened cheese and some Camembert is being made industrially by the ultrafiltration concept. Other countries are also becoming involved.

Cheese yield is dependent upon the fat, protein, and insoluble salt levels in cheesemilk, and upon the amount of water retained in the cheese. The relative contribution of these factors to Cheddar cheese yields is based on classical formulae reported by Van Slyke, which can be applied as well to other cheeses with modification. Yields for the same type of cheese vary from region to region. In Wisconsin, 100 lb of 3.5% fat milk may give 9.6 lb of 37% H_2O Cheddar cheese, but in eastern Canada and northern New York the same weight of 3.5% fat cheese milk may give only 8.9 lb of 37% H_2O cheese. A-natural low casein to fat ratio, caused perhaps by different feeding practices, among a number of factors, exerts a strong influence on the low Cheddar yields of New York and Canada. Supplementing low solids cheesemilk with milk retentates of ultrafiltration provides an opportunity to adjust the fat to casein ratio to optimum, to give maximum yield and highest Cheddar quality. Also, such supplementation increases cheese volume in the same vat, resulting in better economics and greater energy savings. For Cottage cheese made from skim milks supplemented with skim milk retentates to between 3.5% to 5.0% protein to increase volume, there is some preliminary indication also that yield efficiency, actually increases.

Addition of food grade microbial enzymes to cheese curds leads to accelerated ripening and increased flavor of ripened cheese, while immobilized enzyme techniques reported by Olson et al. show a potential for the continuous fermentation of cheese curd without the need for adding milk coagulating enzymes to the cheesemilk.

Historically, cheese has been recognized as a safe food because the lactic acid developed and the many end products of maturation in a ripened cheese, properly made, usually inhibit the growth of food poisoning organisms. Further protection is offered if the cheese milk is pasteurized or heat treated.

In primitive areas, however, where milk often is not pasteurized, where milking animals may be diseased, and where improper cheesemaking is conducted, undulant fever, typhoid fever, and tuberculosis can be contracted through such cheese ingestion. In developed counties, staphylococci and enteropathogenic *E. coli* food poisoning, with less critical effects on mankind, are the more likely causes of illness resulting from natural cheese consumption. Botulism has never been officially reported as resulting from commercial natural cheese in the U.S., but a few outbreaks have been reported as caused by homemade cheeses of insufficient acid development and from commercial process cheese spreads in the U.S. and Argentina.

For some cheeses made inEurope, sodium nitrate supplementation is practiced to suppress innocuous gas-producing spoilage organisms; but residual amounts, reportedly, are very small. More recently, the concept leading to the Bactotherm centrifuge process for continuously removing bacteria spores from milk, heat sterilizing them, and recycling the bactofugate into the cheesemilk has, apparently, enabled European cheesemakers to eliminate nitrate additions in soft cheeses and to reduce nitrate addition by about 75% in ripened cheese.

Some species of molds used to produce foods, including the Penicillium and Aspergillus group, can, under certain circumstances, produce toxins. Although mold spores are used in some ripened cheeses, experimental studies have shown no toxins of mold origin residing in commercial mold ripened cheeses.

Cheeses, in addition to possessing excellent food qualities, possess stimulating flavors which are becoming more appreciated. Flavor development in cheese is extremely complex and variable, depending upon cheese type and quality, environmental conditions, and the presence of critical microorganisms. Such flavor in natural cheese is difficult to duplicate, and it is one of the many reasons why consumption of cheese continues to rise throughout the world.

New Words

buffalo 水牛
precheese 预制干酪
cheddar 契达、干酪
supplement 增补、补充
Camembert 沙门柏干酪
preliminary 初步、开端
Blue mold 青霉干酪
curd 凝块
Brick 砖形干酪

immobilize 使固定、使不动
isoelectric 等电的
coagulate 凝固
rennin 凝乳酶
poison 中毒
postulate 假定
undulant 波浪形的
emulsify 乳化
undulant fever 波型热、米利它热
Roquefort 青纹干酪
typhoid 伤寒(的)
Brie 一种干酪名
tuberculosis 结核病
Penicilliumroquefortior 娄地青霉
staphylococci 葡萄球菌
P. glaucum 灰绿青霉
mycelia 菌丝体
cheesemilk 制干酪用乳
E. coli 大肠埃希氏杆菌
P. caseiocolum 乳酪青霉
botulism 肉毒中毒
P. camemberti 沙门柏干酪青霉
innocuous 无害的,无毒的
proteolytic 分解蛋白质的
centrifuge 离心机
prevail 普遍、盛行
recycle 回收、重新利用
macerated 浸渍,使离析、浸透
bactofugete 细菌离心液
calf 小牛、牛犊
Aspergillus 曲霉属

Exercises

1. What are the fermented milk products? Please write out several fermented milk products.
2. Which factors do have effect on quality of the fermented milk products?
3. What are the benefits of the fermented milk products to people?
4. Where does the cheese come from? How does the cheese come into being?
5. Please write out several roles of cheese.

Unit 3　Meat Techniques of Curing Dry Curing

The object of the curing process is to uniformly distribute the curing ingredients throughout the meat. Of the principal methods in use, the oldest is that of dry curing. In this procedure, the mixture of dry curing ingredients is spread on, or more correctly, rubbed into the surface of the piece of meat to be cured and the meat held under refrigeration until the salt and other materials penetrate to the center.

Penetration rate is dependent on time and temperature. The higher the temperature, the more rapid the penetration. However, temperature much above 40 ℉ will encourage spoilage bacteria and usually result in souring before the cure fully penetrates. Most recommendations give a time-weight relationship for curing. It is generally accepted that it requires a minimum of 40 days for salt concentration in the center of hams to approach the 1% level, regardless of weight. At the same time, the salt concentration in the outer layer of the ham may be 5% to 7%. Penetration of salt from the skin side of the meat is considerably slower than from the lean side. In fact, both skin and fat make relatively effective barriers for cure penetration.

In the case of dry cured meats, the curing mixture is applied to the surface in one, two or three day intervals by a rubbing process. The amount applied is dependent upon final salt concentrations required. Usually, a dry cured product will use upwards of 3% salt based on original green weight.

In dry curing mixtures, the addition of the ascorbates or phosphates is not practiced. Usually, the curing mixture consists of salt, sugar, and nitrate. While nitrite could be added, the curing time is sufficient for nitrate to be broken down to nitrite by bacterial action.

After rubbing with the cure, the products are stacked on shelves or in boxes and held for the prescribed curing time, usually at a temperature of 36 ℉ to 40 ℉. When curing is completed, large pieces, such as hams and shoulders, are held under refrigeration for an additional 30 days to allow the salt concentration to equalize. Following salt equalization, the products are then smoked and aged as desired.

Since most dry cured hams and shoulders are often eaten without sufficient cooking to destroy trichinae, there are rather rigid procedures for handling to insure trichina destruction. This is explained later in this chapter.

Dry curing is sometimes used in conjunction with brine injection for some specialty products. In this procedure, the product is usually pumped with not more than 10% of a fairly strong brine. The balance of the curing ingredients are then applied as a dry rub and the product handled much like dry cured.

Brine Soaking

In terms of historic use, this procedure closely followed dry curing and was used commercially for many years. In brine soaking, the product is placed in a brine solution for an appropriate period of time until the brine penetrates the entire piece. Again, this procedure involves a race against spoilage development and has severe limitations for large pieces, principally because of the relatively slow brine penetrations.

Brine soaking is still used commercially for small items such as tongues, corned beef, etc. brine strength is variable depending on the desired saltiness of the finished product. With this method, there is a tendency to want to reuse brine. This is not a wise practice for several reasons. First of all, the strength of the brine is reduced since some of the curing agents have been incorporated into the meat. Also, the brine at this point has been diluted with meat juices. Finally, the brine has become contaminated with bacteria, something that is absolutely unavoidable.

Brine Injection

Since one of the critical factors involved in curing is the uniform incorporation of the curing ingredients into the meat itself, it would seem logical to assume that forcing the brine directly into the meat would be an advantage. Without brine injection, rapid curing would be unknown.

Stitch or Spray Pumping

With this procedure the brine is pumped into meat with a needle that has a number of holes along its length. Sometimes gangs of needles are employed.

Artery Pumping

In artery pumping, the brine is injected into hams through the femoral artery and follows the arterial system throughout the ham. Proponents of artery pumping ignore the fact that the arterial system in a ham is not uniform in its distribution. The distribution may be sufficient to allow immediate smoking and cooking. However, it is advisable to hold at least 24 hr to allow for uniform cure distribution. Holding in a curing cooler 5 to 7 days after pumping will greatly improve uniformity of cure distribution.

Artery pumping has a disadvantage in that it is a relatively slow procedure and has a high labor requirement. It would not be recommended for an operator who produces more than 400 hams per day.

In order to get a successful job of artery pumping, it requires careful workmanship during slaughter, cutting and subsequent handling to insure that the arteries are intact.

Machine Pumping

The advent of continuous brine injection machines offers the processor an opportunity for rapid processing of hams, bellies, and other cured meat items. The principle used is similar to spray pumping with the exception that the needles inject the brine at hundreds of points. The pressure of the brine forces relatively uniform brine distribution throughout the meat.

It should be remembered that brine localizes in the lean. Overpumping is not a good procedure. The brine forms pockets in the seams between the muscle areas. The result will be the formation of open seams in the finished product.

Care should be taken where brine is recirculated to make periodic checks for dilution and contamination. When the brine shows evidence of accumulation of meat juices, it should be discarded and replaced with fresh brine. The best procedure to follow would be to pump the tank down to the last few inches, discard the balance, and start with fresh brine.

New Words

curing 腌制
corned beef 腌牛肉
uniformly 均匀地
dilute 稀释
rub 擦
contamination 污染
souring 变酸
unavoidable 不可避免的
ham 火腿
logical 合理的
barrier 障碍
stitch 针
interval 间隔
artery 动脉
green weight 毛重
femoral artery 股动脉
ascorbate 抗坏血酸；维生素C
proponent 提倡者
nitrate 硝酸盐
workman ship 工艺
nitrite 亚硝酸盐
slaughter 屠宰
stack 码垛
cutting 分割
prescribed 规定的
intact 完整
shoulder 前腿
advent 降临
age 成熟
overpumping 过度泵入
trichinae 旋毛虫
seam 缝
rigid 严格的
accumulation 积累
injection 注射
discard 丢弃

conjunction with 与……联系
balance 剩余的
brine 卤水、(腌食物用的)盐水
soaking 浸泡

Meat Products

(1) Sectioned and formed products

Various procedures are available for preparing meat products that have the appearance and consistency of intact muscle meat but that are composed of meat pieces that cohere together. These pieces of meat may be fine flakes of frozen and tempered meat or they may be entire muscle systems. The cohesive substance that binds the pieces together can vary from emulsion systems to extracted myofibrillar proteins. Additives that affect the binding strength between particles are salt, phosphates, soy products, dairy powders and enzymatic tenderizers (Schmidt et al, 1981).

Agglomeration of meat particles into a single meat mass or roll by mechanical agitation has been utilized for several years. The agitation or tumbling of meat particles in the presence of a salt solution brings the salt soluble proteins to the meat surface. These proteins coagulate during heating to bond the meat surfaces. Salt aids meat particle binding and reduces fluid loss during heating. Alkaline phosphates significantly increase the binding strength of sectioned and formed meat products. Binding strength also improves with an increase in the internal cooking temperature. Greater cooking yields and binding strength occur when the meat is treated both with salt and phosphate rather than either alone. Both pre-and post-rigor materials can be utilized to produce sectioned and formed meat products. Generally, meat that is boned pre-rigor and subjected to mechanical action with the addition of salt and phosphate has greater protein extraction during the treatment. The increased protein extraction will aid in particle-to-particle cohesion in the finished product.

The particles that are utilized to produce sectioned and formed products may be entire muscles, very coarsely ground meat or flaked meat. Large sections may be produced by cutting muscle chunks into sections by hand or by using a dicer. Some particles can be produced by using a plate in a meat grinder that has larger kidney-shaped holes in it. Particles of various sizes can be produced by using a flaking machine that is capable of varying the flake size from very fine to coarsely flaked material. It is possible to incorporate a certain amount of mechanically deboned tissue or finely chopped material into the sectioned and formed product. If additional fat or connective tissue must be incorporated into the material, it is wise to incorporate it as a finely chopped emulsion for inclusion in the final product.

The mechanical energy that must be applied to the various size particles and other ingredients to extract myofibrillar proteins can be provided by a mixer, tumbler or massager. Generally, when particles are of a relatively small size, they can be combined with other ingredients and subjected to mechanical action in either a paddle or ribbon mixer or blender. There are mixers and blenders on the market that have specially designed paddles that will not tear up larger chunks of meat. Whether a mixer or tumbler is used, it has been found useful to vacuumize the treatment. By the application of a vacuum during mixing, tumbling or massaging, the exudates that is produced will not foam. An exudate that has not foamed provides a stronger bond of more uniform appearance between the meat particles.

Tumbling generally refers to placing meat inside a stainless steel drum that rotates at such a speed that some of the meat is carried to the top of the drum and drops down at least 1m onto the bottom of the drum. This impact of meat on meat, as well as the friction of one portion abrading another, has several functions: it aids in abrading the myofibrillar proteins from the surface of the meat, it makes the meat more pliable and increases the rate of cure distribution. Many of the tumblers operate on an intermittent tumbling cycle such that the product is subjected to tumbling followed by a rest period. Some systems use needles or bars to abrade and penetrate the surface of the meat so as to encourage absorption of the brine and scrape bits of muscle into the liquid. Some tumblers are equipped with simultaneous injection devices for injecting brine during the tumbling operation.

Massaging is generally a less severe treatment than tumbling. Massagers come in many sizes and designs. Most models use a bin similar to a standard meat vat which is equipped with a large motor to power a vertical shaft that has arms attached to it. The massager slowly stirs the large chunks of meat. A mixer will do the job very well for smaller pieces of meat that have been ground.

Regardless of the system used, it is important that the machine manipulates all the pieces of meat equally. In addition, it is recommended that the first manipulation takes place shortly after the meat has been injected with pickle or the pickle has been added to the mixer. For development of color and uniformity of distribution, it is sometimes recommended that a period of approx. $10 \sim 24$ h be allowed between the initial and final mechanical action. Shortly after the final mechanical action has been applied, it is recommended that the material be loaded into its final mold whether it be a metal mold, casing or can. This is done shortly after the final mechanical action because the tissue remains pliable for a short period of time. This will permit the tissue to fit the mold more uniformly and results in fewer voids in the finished product.

Some of the advantages of tumbling and massaging are as follows:

(1) The processor is able to create products of desired shape, weight and sliceability.

(2) Meat becomes pliable and is easily handled mechanically.

(3) The consumer is presented with a product of uniform composition.

(4) Cooking losses can be controlled.

(5) Muscle tissues from many species and many parts of the anatomy that formerly were unacceptable for roasts and steaks can be separated and the appropriate portions made acceptable for inclusion in the sectioned and formed roasts and steaks.

(6) Tenderness can be increased.

New Words

section　切块

tear up 撕裂
consistency 质地
vacuumize 使真空化
intact 完整的
exudates 渗出液
cohere 粘接
drum 圆桶
phosphate 磷酸盐
impact 冲击
tenderizer 嫩化剂
friction 摩擦
agglomeration 聚集
abrade 摩擦
agitation 搅拌
pliable 有柔韧性的
tumbling 滚打
intermittent 间歇的
alkaline 盐溶性
penetrate 穿刺
significantly 碱性的
scrape 刮
pre-rigor 死僵前
bit 碎肉片
post-rigor 死僵后
simultaneous 同步的
bone 剔骨
bin 箱
coarsely 粗糙的
vertical 垂直的
chunk 块
manipulate 操纵
dicer 方粒切粒机
pickle 腌制剂
grinder 绞肉机
uniformity 均匀性
deboned 去骨的
approx = approximately 大约
chop 斩拌
void 空隙

mixer 拌馅机
sliceability 切片性
tumbler 滚打机
anatomy 解剖学
massager 按摩机
roast 烤肉
blender 混合机
steak 牛排
tenderness 嫩度
appropriate 合理的

(2) Sausage cooked after formulation

The mincer, bowl chopper, vacuum mixer, blender, mill and stuffer are the machines most used. Small quantities are manufactured in batches, while large quantities are prepared semi-continuously. The preblending system is gaining popularity in Europe. Although originally a means to obtain uniform raw material, the combination of mincer, preblender, mill and stuffer now provides a more or less continuous production system. For critical products the conventional chopper mixer stuffer process is preferred. Many types of casings are used, but in central Europe many small sausages are produced in natural casing.

Pork and beef are the most popular meat ingredients, with limited use of mutton, horse and poultry. Variety meats such as hearts, diaphragm, tripe, pork skin, stomachs, etc. are allowed in most countries, although not in every product. In some countries, sausage products are divided in two or three quality categories. For each of them the raw materials which can be used are defined. Mechanically deboned meat is used in an increasing number of countries. In some countries the amounts are limited by law, in other countries by technological or sensoric considerations.

The use of all adipose tissue in any form is allowed: as whole cuts, such as pork bellies, or as fat trimmings. The successful production of cooked sausage depends, among other things, on the type of fat used. Firm adipose tissue such as pork-back fat, belly fat, jowl and ham fat are preferred for different applications. The European sausage manufactures use all kinds of fat in any cooked sausage which is not the case in the American meat industry. Europeans produce stable cooked sausage of good quality with any type of animal fat by pre-emulsifying the fat in water by means of a non-meat protein such as soy or milk. This process became known and popular about 20 years ago and is still widely used today.

In many factories, all kinds of fat trimmings and other fatty tissue are emulsified. The fresh or chilled fat is chopped, the caseinate is added, followed by the addition of boiling water. After 4~6 min chopping, 1.5%~2% salt is added. After another minute of chopping, the emulsion is taken out of the chopper and chilled. After chilling, the emulsion can be incorporated in almost any type of cooked sausage. For optimal emulsion stability, chopping time and emulsification temperature are of great importance. For some purposes, emulsification can be performed with the use of cold water; or the hot emulsion, made with a part of the water, may be cooled with the balance made up by ice.

This method is used for cooked meat sausage and loaves and enables direct incorporation of the emulsion into the sausage mix.

About one-third of the total fat in the formula can be used in preemulsified form. This method provides better fat and water binding, greater heat stability, an absence of any greasy taste and a production process which is considerably less sensitive to chopping time and emulsion temperature.

The ingredients of cooked sausage may vary from pure lean meat, ice, fat, curing salts and spices to sausage consisting of as little as 20% meat and considerable amounts of rinds, water, fat, additives and extenders. Fat contents range from 15% to over 35%. Collagen-rich materials vary from none to more than 15% rinds. Extenders and additives which promote fat and water binding, such as polyphosphates, range from less than 0.3% polyphosphate or 2% sodium caseinate to almost unlimited use of all possible extenders, and in addition, up to 0.5% polyphosphate. The legal regulations have a great influence on the ingredients of the sausage. Where the authorities are generous with respect to added water but forbid most of the extenders and polyphosphate, sausage making still is an art and recipe variations are limited. When the use of connective tissue is restricted, the sausage products are of excellent quality but also expensive.

Where added water is forbidden, the possibilities of producing a good quality sausage are rather low and the manufacture of lower-priced sausage necessitates the extensive use of collagen-rich material, often combined with a high fat content. In countries where added water is not regulated and extenders are allowed, lean meat content is generally low, whereas the amount of extenders and water content are high.

Most of the extenders used are starches and flours (potato, corn, wheat), milk proteins (sodium caseinate), skimmed milk powder, soy products (isolate and concentrate), wheat gluten, egg white, and blood plasma, either frozen or dried.

Other additives used are water, salt, phosphates, citrates, ascorbic acid, nitrate, nitrite, monosodium glutamate, glucono delta lactone, dextrose, malt dextrin, lactose and smoke concentrates. The most popular spices are pepper, nutmeg, coriander, ginger, mace, cardamom, paprika and garlic.

New Words

mincer　绞肉机
rind　筋腱
stuffer　灌肠机
extender　黏结剂
batch　小批量
collagen-rich　富含胶蛋白的
semi-continuously　半连续的
promote　促进
preblending　预混合

polyphosphate 聚磷酸盐
conventional 传统的
sodium caseinate 酪朊酸钠
casing 肠衣
legal regulation 法规
natural casing 天然肠衣
recipe 配方
pork 猪肉
gluten 面筋
mutton 羊肉
plasma 浆
poultry 禽
citrate 柠檬酸盐
diaphragm 横膈
paprika 红椒
tripe 牛肚
garlic 大蒜
stomach 肚
nitrite 亚硝酸盐
category 分类
monosodium glutamate 单谷氨酸钠
adipose 脂肪的
glucono-delta-lactone 葡萄糖酸-δ-内酯
belly 腹肉
dextrose 葡萄糖
trimming 边角料
malt dextrin 麦芽糊精
pork-back fat 猪背脂
lactose 乳糖
jowl 颊肉
smoke concentrate 浓缩烟熏剂
ham 火腿
pepper 胡椒
caseinate 酪蛋白盐
nutmeg 豆蔻
meat loaf 肉糕
coriander 胡荽、香菜
formula 配方
cardamom 小豆蔻

greasy   油腻的
ginger   姜
lean   瘦肉
mace   肉豆蔻
curing   腌制
spice   香料

Exercises

1. What are the aims of dry curing to meats?
2. Please speak out the process of dry curing for meats.
3. What are the differents between Brine soaking and Brine Injection?
4. What is the process of Brine Injection?
5. What are the advantages of tumbling and massaging?

Unit 4　Eggs

Foreword

Eggs have been a human food since ancient times. They are one of nature's most nearly perfect protein foods and have other high quality nutrients. Eggs are readily digested and can provide a significant portion of the nutrients required daily for growth and maintenance of body tissues. They are utilized in many ways both in the food industry and the home. Chicken eggs are the most important. Those of other birds (geese, ducks, plovers, seagulls, quail) are of lesser significance. Thus, the term "eggs", without a prefix, generally relates to chicken eggs and is so considered in this chapter.

Structure, Physical Properties and Composition

The egg is surrounded by a 0.2~0.4 mm thick calcareous and porous shell. Shells of chicken eggs are white-yellow to brown, duck's are greenish to white, and those of most wild birds are characteristically spotted. The inside of the shell is lined with two closely-adhering membranes (inner and outer). The two membranes separate at the large end of the egg to form an air space, the so-called air cell. The air cell is approx. 5 mm in diameter in fresh eggs and increases in size during storage, hence it can be used to determine the age of eggs. The egg white (albumen) is an aqueous, faintly straw-tinted, gel-like liquid, consisting of four fractions that differ in viscosity. The inner portion of the egg, the yolk, is surrounded by albumen. A thin but very firm layer of albumen (chalaziferous layer) closely surrounds the yolk and it branches on opposite sides of the yolk into two chalazae that extend into the thick albumen. The chalazae resemble two twisted rope-like cords, twisted clockwise at the large end of the egg and counter-clockwise at the small end. They serve as anchors to keep the yolk in the center. In an opened egg the chalazae remain with the yolk. The germinal disc (blastoderm) is located at the top of a club-shaped latebra on one side of the yolk. The yolk consists of alternate layers of dark-and light-colored material arranged concentrically.

Storage of Eggs

A series of changes occurs in eggs during storage. The diffusion of CO_2 through the pores of the shell, which starts soon after the egg is laid, causes a sharp rise in pH, especially in egg white. The gradual evaporation of water through the shell causes a decrease in density (initially approx. 1.086 g/cm^3) and the air cell enlarges. The viscosity of the egg white drops. The yolk is compact and upright in a fresh egg, but it flattens during storage. After the egg is cracked and the contents are released onto a level surface, this flattening is expressed as yolk index, the ratio of yolk height to diameter. Furthermore, the vitellin membrane of the yolk becomes rigid and tears readily once the egg is opened. Of importance for egg processing is the fact that several properties change, such as egg white whipping behavior and foam stability. In addition, a "stale" flavor develops.

These changes are used for determination of the age of an egg, e.g. floating test (change in egg density), flash candling (egg yolk form and position), egg white viscosity test, measurement of air cell size, refractive index, and sensory assay of the "stale" flavor (performed mostly with soft-boiled eggs). The quality loss during storage of eggs is lower as the storage temperature is lower, as are the losses of CO_2 and water. Therefore, cold storage is an important part of egg preservation. A temperature of 0℃ to -1.5℃ (common chilled storage or subcooling at -1.5℃) and a relative humidity of 85%~90% are generally used. A coating (oiling) of the shell surface with light paraffin-base mineral oil quite efficiently retards CO_2 and vapor escape, but a tangible benefit is derived only if oil is applied soon (1h) after laying, since at this time the CO_2 loss is the highest. Controlled atmosphere storage of eggs (air or nitrogen with up to 45% CO_2) has been shown to be a beneficial form of egg preservation. Cold storage preserves eggs for 6~9 months, with a particularly increased shelf life with subcooled storage at -1.5℃. Egg weight loss is 3.0%~6.5% during storage.

Egg Products

Egg products, in liquid, frozen or dried forms, are made from whole eggs, white or yolk. They are utilized further as semi-end products in the manufacturing of baked goods, noodles, confectionery, pastry products, mayonnaise and other salad dressing, soup powders, margarine, meat products, ice creams and egg liqueurs.

New Words

plover 鸻科
blastoderm 胚盘
seagull 海鸥

club   棒
quail   鹌鹑
latebra   蛋黄心
bird   鸟
coefficient   导数
calcareous   钙质
vitelline   蛋黄的
porous   多孔的
rigid   刚性的
tangible   明确的
stale   腐败的
purification   纯化
coat   涂膜
confectionery   蜜饯
paraffin   石蜡
salad   色拉
pore   气孔
margarine   人造奶油
spotted   斑点
cement   粘合
adhere   胶着
mammilla   乳头
aqueous   水溶液
spongy   海绵
strawtint   淡黄色的
anchor   锚
viscosity   黏度
germinal   萌发的
albumen   蛋白

Exercises

1. What are the Structure, Physical Properties and Composition of eggs?
2. How to store eggs?
3. How to determine the age of an egg?

References

1. Food safety and quality assurance issues-*Improving the Safety and Quality of Fresh Fruits and Vegetables: A Training Manual for Trainers*. 2002
2. National Research Council (NRC). 1998. *Ensuring Safe Food: From Production to Consumption*. Washington: National Academy Press.
3. Potter, Morris E. 1996. Factors for the Emergence of Foodborne Disease. In *Proceedings of the Fourth ASEPT International Conference, Food Safety* 1996. 185–195.
4. Tauxe, R., H. Kruse, C. Hedberg, M. Potter, J. Madden, and K. Waschsmuth. 1997. Microbial Hazards and Emerging Issues Associated with Produce. A preliminary report to the National Advisory Committee on Microbiologic Criteria for Foods. *Journal of Food Protection* 60(11): 1400–1408.
5. Food Quality and Safety Systems—A Training Manual on Food Hygiene and the Hazard Analysis and Critical Control Point (HACCP) System. Food Quality and Standards Service Food and Nutrition Division, FOOD AND AGRICULTURE ORGANIZATION OF THE UNITED NATIONS Rome, 1998.
6. Foster, E. M. 1997. Historical overview of key issues in food safety. Emerg. Infect. Dis. 3: 481–482.
7. Thorne, Stuart. 1986. *The History of Food Preservation*. Totowa, NJ: Barnes and Noble Books.
8. Mottram D S, Wedzicha B L, Dodson A T. Acrylamide is formed in the Maillardreaction [J]. Nature, 2002, 419: 448–449
9. Stadler R H, Blank I, Varge N, et al. Acrylamide from Maillard reaction products [J]. Nature, 2002, 419: 449–450
10. FAO/WHO, Summary and Conclusions of Joint FAO/WHO committee on food additives sixty-fourth meeting Rome, 8-17 February 2005, JECFA/64/SC.
11. European Food safety Authority, Statements of the scientific panel on contaminants in the food chain a summary report on acrylamide in food of the 64th meeting of the Joint FAO/WHO expert committee on food additives, adopted on 19 april 2005.
12. Tareke E, Rydberg P, Karlsson P, et al. Analysis of acrylamide, acarcinogen formed in heated foodstuffs[J]. J Agri Food Chem, 2002, 50: 4998–5006.
13. Becalski A, Lau P2YB, Lewis D, et al. Acrylamide in foods: occurrence, sources, and modeling[J]. J Agri Food Chem, 2003, 51: 802–808.
14. Yaylayan V A, Wnorowski A, Locas C P. Why asparagine needs carbohydrate to generate acrylamide[J]. J Agri Food Chem, 2003, 51: 1753–1757.
15. Food allergy. http://www.emedicinehealth.com/food_allergy/article_em.htm.
16. The food allergy and anaphylaxis network. http://www.foodallergy.org/allergens/index.html.
17. Allergy and Allergies. http://www.ei-resource.org/allergies.asp.

18. Genetically Modified Foods: Harmful or Helpful? http://www.csa.com/discoveryguides/gmfood/overview.php.

19. Genetically Modified Foods and Organisms. http://www.ornl.gov/sci/techresources/Human_Genome/elsi/gmfood.shtml.

20. 50 Harmful Effects of Genetically Modified Foods. http://www.cqs.com/50harm.htm.

21. MAD COW DISEASE, the BSE Epidemic in Great Britain. http://www.accessexcellence.org/WN/NM/madcow96.html.

22. Mad Cow DiseaseOverview. http://www.webmd.com/a-to-z-guides/Mad-Cow-Disease-Overview.

23. Mad Cow Disease What the Government Isn't Telling You! http://www.drday.com/madcow.htm.

24. Mad Cow Disease and Humans. http://rarediseases.about.com/od/rarediseases1/a/vcjd.htm.

25. Food Irradiation: A Safe Measure. http://www.fda.gov/opacom/catalog/irradbro.html.

26. Food Irradiation. http://www.physics.isu.edu/radinf/food.htm.

27. Food Safety and Irradiation: Protecting the Public from Foodborne Infections. http://www.cdc.gov/ncidod/eid/vol7no3_supp/tauxe.htm.

28. By Brian P. Baker, Charles M. Benbrook, Edward Groth III, and Karen Lutz Benbrook. Published in: Food Additives and Contaminants, Volume 19, No. 5, May 2002, pages 427–446.

29. Pesticide Residues In Foods: Is Food Safety Just A Matter Of Organic Versus Traditional Farming? http://www.uky.edu/Ag/Entomology/entfacts/misc/ef009.htm.

30. An Analysis Of U.S. Government Data On Pesticide Residues In Foods. http://www.consumersunion.org/food/do_you_know2.htm.

31. Code of Federal Regulations. Part 110—Current Good Manufacturing Practice In Manufacturing, Packing, Or Holding Human Food.

32. *cGMPs/ Food Plant Sanitation*, Wilbur A. Gould, Ph.D., CTI Publications, 1990.

33. Codex Alimentarius Commission (CAC) Recommended Code of practice. 2003. Recommended international code of practice general principles of food hygiene. CAC/RCP 1–1969, Rev. 4–20031.

34. Food and drug administration The Good Manufacturing Practice (GMP-Quality System Regulation) Final Rule http://www.fda.gov/cdrh/comp/gmp.html.

35. National Meat Association coordinated by Institute of Food Science and Engineering. Texas A&M University College Station, Texas. 1998. Guidelines for Developing Good Manufacturing Practices (GMPs) and Standard Operating Procedures (SOPs) for Raw Ground Products http://haccpalliance.org/alliance/Nmagmp2.pdf.

36. Ken Hilderbrand, Oregon State University. 1996. Sanitation Standard Operating Procedures. http://www-seafood.ucdavis.edu/haccp/ssop/lox.htm.

37. Model Sanitation Standard Operating Procedures. http://www-seafood.ucdavis.edu/

haccp/ssop/ssop2. htm.
38. U. S. Food and Drug Administration. HACCP. http://www. cfsan. fda. gov/~lrd/haccp. html.
39. Hazard Analysis and Critical Control Point (HACCP) System and Guidelines For Its Application. http://www. fao. org/DOCREP/005/Y1579E/y1579e03. htm.
40. USDA, (1999), Guidedbook for the preparation of HACCP Plans, U. S. Department of Agriculture Food and Safety Inspection Service Document HACCP-1.
41. Hazard Analysis Critical Control Point (HACCP). http://www1. agric. gov. ab. ca/ $department/deptdocs. nsf/all/afs4338.
42. Anonymous. *Hazard Analysis Critical Control Point System*. United States National Advisory Committee on Microbiological Criteria for Foods. United States Department of Agriculture, Washington DC, 1992.
43. Anonymous. *A Guide to the Implementation of Hazard Analysis Critical Control Point Systems in the Seafood Industry*. FIICC/MAF Regulatory Authority (Meat & Seafood) Wellington, 1994.
44. National Restaurant Assoc. The Education Foundation Publication. 1993. HACCP Reference Book.
45. FDA, DHHS publication, 2nd ed. 1997. HACCP-Regulatory Applications in Retail Food Establishments.
46. The University of Arizona fact sheet, 1997. "Hazard Analysis Critical Control Points (HACCP) for the Consumer".
47. Food Safety Risk Analysis. http://depts. washington. edu/foodrisk/index. html.
48. Definitions of risk analysis terms related to food safety. http://www. who. int/foodsafety/publications/micro/riskanalysis_definitions/en/.
49. Overview of Risk Analysis. http://www. agnr. umd. edu/fsra/workshop. htm.
50. These decisions include the *Statements of principle concerning the role of science in the Codex decision-making process and the extent to which other factors are taken into account* and the *Statements of principle relating to the role of food safety risk assessment* (Codex Alimentarius Commission Procedural Manual; Thirteenth edition).
51. FAO and WHO, 1999a. The Application of Risk Communication to Food Standards and Safety Matters. Report of the joint FAO/WHO expert consultation. Rome, Italy, 2-6 February 1998. FAO, Rome.
52. FAO/WHO. In press. Exposure assessment of microbiological hazards in foods: Guidelines. Microbiological Risk Assessment Series, No. 7.
53. Byrd, D. M. & Cothern, C. R. 2000. Introduction to risk analysis. ABS Consulting, Government Institutes Division, Rockville, Maryland.
54. Vose, D. 2002. Risk analysis: a quantitative guide. Second edition. John Wiley and Sons, New York.
55. The Presidential/Congressional Commission on risk assessment and risk management. 1997. Final report, Volume 2.